Scientific American's

ASK THE EXPERTS

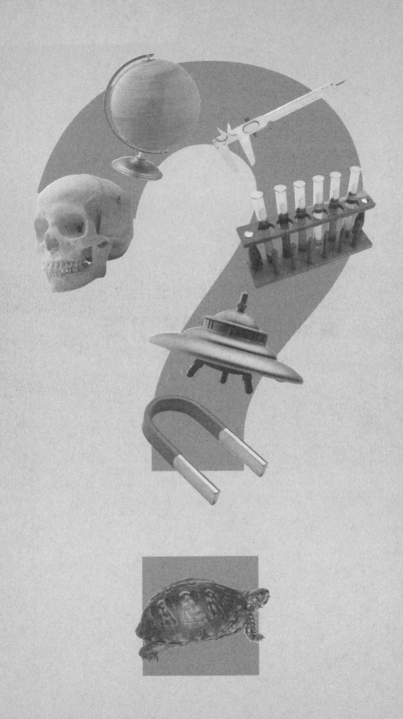

Answers to the Most
Puzzling and Mind-Blowing
Science Questions

Scientific American's
ASK THE EXPERTS

by the editors of **SCIENTIFIC AMERICAN**

HarperResource
An Imprint of HarperCollinsPublishers

HarperCollins books may be purchased for educational, business, or
sales promotional use. For information please write: Special Markets
Department, HarperCollins Publishers Inc., 10 East 53rd Street,
New York, NY 10022.

FIRST EDITION

Designed by William Ruoto

Library of Congress Cataloging-in-Publication Data
has been applied for.

ISBN 0-06-052336-0

03 04 05 06 07 WBC/RRD 10 9 8 7 6 5 4 3 2

Contents

1 Celestial Bodies
ASTRONOMY

It Came From Outer Space
asteroids, meteors, and comets

Heavenly Bodies
planets and moons

Star Light, Star Bright
stars

Far, Far Away . . .
the universe

It's Alive!
BIOLOGY

That's a Horse of a Different Color
animal kingdom

Talkin' About Evolution
evolution

When Dinosaurs Ruled the Earth
dinosaurs

3   # Being Human

It's All in the Genes
human evolution

Oh, Behave!
human behavior

You Haven't Aged a Bit
growing older

Anatomy 101
the human body

The Dr. Is In
health and medicine

4 As a Matter of Fact

CHEMISTRY

Elementary, My Dear Watson . . .
the elements

If You Can't Stand the Heat, Get Out of the Kitchen!
everyday chemistry

Where There's Smoke, There's a Fire
more chemistry

5 There's No Place Like Home
EARTH SCIENCE

Everybody Talks About It . . .
weather

Up Above
the atmosphere

The Upper Crust
earth's surface and below

Let's Get Wet
oceans

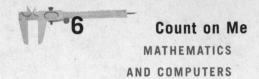

6   Count on Me

MATHEMATICS
AND COMPUTERS

7 Let's Get Physical
PHYSICS

Bottom of the 9th, Bases Loaded
the physics of baseball

Scientific American's
ASK THE EXPERTS

Celestial Bodies

It Came From Outer Space
asteroids, meteors, and comets

?

How crowded is the asteroid belt?

ANSWERED BY:

Tom Gehrel, University of Arizona, Tuscon, Arizona.
A veteran asteroid hunter, he and his colleagues find
roughly 20,000 objects a year—many of them
uncatalogued asteroids—using the Spacewatch
Telescope on Kitt Peak.

S ome scientists were seriously concerned about the possible high density of objects in the asteroid belt, which lies between the orbits of Mars and Jupiter, when the first robotic spacecraft were scheduled to be sent through it. The first crossing of the asteroid belt took place in the early 1970s, when the *Pioneer 10* and *Pioneer 11* spacecraft journeyed to Jupiter and beyond. The danger does not lie in the risk of hitting a large object. In fact, such a risk is minuscule because there is a tremendous amount of space between Mars and Jupiter and because the objects there are very small in relation. Even though there are perhaps a million asteroids larger than one kilometer in diameter, the chance of a spacecraft not getting through the asteroid belt is negligible.

Even if there were 100,000 sizable asteroids (more than a few kilometers in size) in the asteroid belt—and the real number is quite likely about 10 times less—the average separation between them would be about five million kilometers. That is more than 10 times the distance between the earth and the moon. If you were standing on one of those asteroids and looked up, you would not see a sky full of asteroids; your neighbors would appear so small and dim that you would be quite lucky to even see one, let alone hundreds.

In some ways, the asteroid belt is actually emptier than we might like. In the early 1990s, the National Aeronautics and Space Administration wanted the *Galileo* spacecraft to encounter an asteroid while it was passing through the asteroid belt on its way to Jupiter. But it took some effort to find an object that was located even roughly along *Galileo*'s path. Special targeting was required to reach this object, but the result was the first close-up view of an asteroid, the one called Gaspra.

The number of objects in the asteroid belt increases steeply with decreasing size, but even at micrometer sizes the *Pioneer* spacecraft were hit only a few times during their passage. That is not to say that asteroids cannot pose any danger, however. It is worth noting that for a large planet like Earth, over a long period of time, there is an appreciable chance of being hit. This hazard comes from the fragments of mutual collisions in the asteroid belt; after their break-up, some of these fragments move toward the earth under the gravitational action of Jupiter.

An asteroid about 12 kilometers in diameter crashed into the earth 65 million years ago, killing nearly 90 percent of the animals, including the dinosaurs. Such major impacts are very rare events, but for smaller objects the likelihood of impact increases; the chance of the earth being struck by an object approximately one kilometer in size is about one in 5,000 in a human lifetime. An object one kilometer across would still be large enough to cause a global disaster because of the enormous energy it would release upon impact: at least a million times the energy of the bomb dropped on Hiroshima in 1945.

?

What causes a meteor shower?

ANSWERED BY:
Gregory A. Lyzenga, Professor of Physics,
Harvey Mudd College, Claremont, California.

Meteor showers occur when the earth in its orbit around the sun passes through debris left over from the disintegration of comets. Although the earth's orbit around the sun is almost circular, most comets travel in orbits that are highly elongated ellipses. As a result, some comets have orbits that intersect or partially overlap the earth's path.

Because a comet's nucleus is made up of a combination of icy materials and loosely consolidated "dirt," when a comet is heated by passing close to the sun, it more or less slowly disintegrates, producing the visible tail. The rocky debris, consisting of mostly sand-size particles, continues in an elongated orbit around the sun close to that of its parent comet. When the earth intersects this orbit in its annual trip, it can run into this debris, which burns up on entry into the earth's atmosphere, producing a visible shower of meteors.

Meteor showers associated with particular comet orbits occur at about the same time each year, because it is at those points in the earth's orbit that the collisions occur. However, because some parts of the comet's path are richer in debris than others, the strength of a meteor shower may vary from one year to the next. Typically a meteor shower will be strongest when the earth crosses the comet's path shortly after the parent comet has passed.

?

Is it possible that a meteorite could strike a commercial airliner and cause it to explode?

ANSWERED BY:
**David Morrison, NASA Ames Research Center,
Maffett Field, California.**

I t is certainly possible, although the probability is low. We can make a very rough estimate by comparing the area of airliners with the area of cars in the United States. A typical car has an area on the order of 10 square meters, and there are roughly 100 million cars in the United States, for a total area of about 1,000 square kilometers. The typical airliner has a cross-sectional area of several hundred square meters, but the number of planes is much smaller than the number of cars, perhaps a few thousand.

The total area of airliners is therefore no more than 10 square kilometers, or a factor of at least 100 less than that of cars. Three cars are known to have been struck by meteorites in the United States during the past century, so it would appear that the odds are against any airplanes having been hit, but it is not impossible that one might have been.

If an airplane were hit, it would be more likely to occur on the ground than in the air, because airplanes spend more time overall on the ground.

?

Why are impact craters always round?

ANSWERED BY:

Gregory A. Lyzenga, Professor of Physics, Harvey Mudd College, Claremont, California.

When geologists and astronomers first recognized that craters were produced by impacts, they surmised that much of the impacting body might be found still buried beneath the surface of the crater floor. Much later, however, scientists realized that at typical solar system velocities—several to tens of kilometers per second—any impacting body must be completely vaporized when it hits an object.

At the moment an asteroid collides with a planet, there is an explosive release of the asteroid's huge kinetic energy. The energy is very abruptly deposited at what amounts to a single point in the planet's crust. This sudden, focused release resembles more than anything else the detonation of an extremely powerful bomb. As in the case of a bomb explosion, the shape of the resulting crater is round: Ejecta are thrown equally in all directions regardless of the direction from which the bomb may have arrived.

This behavior may seem at odds with our daily experience of throwing rocks into a sandbox or mud, because in those cases the shape and size of the "crater" is dominated by the physical dimensions of the impactor. In the case of astronomical impacts, though, the physical shape and direction of approach of the meteorite is insignificant compared with the tremendous kinetic energy that it carries.

An exception to this rule occurs only if the impact occurs at an extremely shallow, grazing angle. If the angle of impact is quite close to horizontal, the bottom, middle, and top parts of the impacting asteroid will strike the surface at separate points spread out along a line. In this case, instead of the energy being deposited at a point, it will be released in an elongated zone—as if our "bomb" had the shape of a long rod. This requires an impact at an angle of no more than a few degrees from horizontal. For this reason, the vast majority of impacts produce round or nearly round craters, just as is observed.

Heavenly Bodies
planets and moons

?

What defines a true planet, and why might Pluto not qualify?

ANSWERED BY:
Daniel W. E. Green, Harvard-Smithsonian Center for Astrophysics, Cambridge, Massachusetts.

Anything in the solar system that is larger than a few meters in size and that does not produce stellar quantities of heat and light is properly considered a planet of some sort (major planet, minor planet, small planet, tiny planet), unless it orbits another body besides the sun—in which case it is usually called a satellite of the larger body. Though not usually called planets, even comets could be thought of as small, icy planets. For this reason, Pluto is surely a planet.

There are different kinds of planets in the solar system, and we cannot adequately classify all of them, because we do not have enough information. A large group of planets 1,000 kilometers across and smaller orbit the sun in a large belt between the orbits of Mars and Jupiter; these small planets are usually referred to as asteroids or minor planets. Most asteroids have orbits that keep them always between Mars and Jupiter (that is, their orbits are not terribly elongated). Many of them, however, have orbits that take them across the orbits of major planets, including that of Earth. Comets are well known to have extremely elongated, or highly elliptical, orbits. Some comets, such as Halley's comet, have orbits that cross the orbits of many of the eight major planets.

Eight major planets, you might ask? Well, in reality, it has come time to stop talking generally about the number of planets in the solar system, because such figures can be misleading. It is probably better to say simply that the solar system is a system of objects that is dominated by a star (the sun), around which are found many orbiting bodies of sizes ranging from particles of dust and gas to the giant gaseous planet Jupiter. There are four notable large gaseous planets (Jupiter, Saturn, Uranus, and Neptune), and inside the main asteroid belt are some smaller, rocky bodies (the most sizable of which are Mercury, Venus, Earth, and Mars). Orbiting the large gaseous planets are dozens of satellites, seven of which, along with our own moon, are larger than Pluto.

It is only since 1978 that we have known the real size and approximate mass of Pluto; both of these are far smaller than astronomers had thought soon after Pluto's discovery in 1930. For numerous reasons, Pluto was called the "ninth planet" at the time; with little information to support or refute that assertion, this classification became locked in astronomy textbooks for decades. But all along, Pluto appeared different from the eight known larger planets: For instance, its orbit is much more elliptical and more highly inclined with respect to the ecliptic than are the orbits of the larger planets, and its orbit brings it inside that of Neptune, so that Neptune is currently the outermost major planet. Pluto is so small that calling it a "major planet" is misleading in the context of what we now know about the solar system. It is more accurately described as a "planetesimal" or a "minor planet." There is even evidence that Pluto may in effect be a giant comet. But much more work and observation is needed before drawing clear conclusions.

?

Why do the moon and the sun look so much larger near the horizon?

ANSWERED BY:

Maurice Hershenson, Professor of Psychology, Brandeis University, Waltham, Massachusetts.

The so-called "moon illusion" is one of the oldest known psychological phenomena; records of it go back to ancient China and Egypt. It may be the most ancient scientific puzzle that is still unexplained.

People trained in the physical sciences often think that the illusion is real, that the moon actually looks large when it is near the horizon because of the refraction of light by the atmosphere. In fact, there is a very small refractive effect, but it is not the cause of the illusion.

There are a couple of ways you can prove to yourself that the light reaching the eye from the moon remains the same as the moon changes position in the sky. For instance, if you photograph the moon at various heights above the horizon, you will see that the images of the moon are all the same size. My students frequently send me photos of a "giant" harvest moon in which the moon looks like a small spot in the sky. (The same thing happens in photos of seemingly spectacular sunsets—the illusion works for the sun as well.) Another way to break the hold of the illusion is to cup your hand into a fist and look through it at the "large" horizon moon. It will immediately shrink in size. Clearly, this is a psychological effect.

My own view is that the moon illusion is linked to the mechanism that produces everyday size–distance perception, a genetically determined brain process that allows us to translate the planar images that fall on the retina into a view of rigid objects moving in space. I believe the moon illusion results from what happens when the mechanism operates in an unusual situation. In normal perception, when rigid objects move in depth (distance), the angular size of the light image stimulating our eyes grows or shrinks. The brain automatically translates this changing stimulation back into the perception of rigid objects whose position in depth is changing.

When the moon is near the horizon, the ground and horizon make the moon appear relatively close. Because the moon is changing its apparent position in depth while the light stimulus remains constant, the brain's size–distance mechanism changes its perceived size and makes the moon appear very large.

?

What is a blue moon?

ANSWERED BY:

George F. Spagna, Jr., Chair, Department of Physics,
Randolph-Macon College, Ashland, Virginia.

The definition has varied over the years. A blue moon once meant something virtually impossible, as in the expression "When pigs fly!" This was apparently the usage as early as the sixteenth century. Then, in 1883, the explosion of the volcano Krakatau in Indonesia released enough dust to turn sunsets green worldwide and the moon blue. Forest fires, severe drought, and volcanic eruptions can still do this. So a blue moon became synonymous with something rare—hence the phrase "once in a blue moon."

The more recent connection of a blue moon with the calendar apparently comes from the 1937 *Maine Farmer's Almanac.* The almanac relies on the tropical year, which runs from winter solstice to winter solstice. In it, the seasons are not identical in length and the earth's orbit is elliptical. Further, the synodic, or lunar, month is about 29.5 days, which doesn't fit evenly into a 365.24-day tropical year or into seasons roughly three months in length.

Most tropical years have 12 full moons, but occasionally there are 13, so one of the seasons will have 4. The almanac called that fourth full moon in a season a blue moon. (The full moons closest to the equinoxes and solstices already have traditional names.) J. Hugh Pruett, writing in 1946 in *Sky and Telescope,* misinterpreted the almanac to mean the second full moon in a given month. That version was repeated in a 1980 broadcast of National Public Radio's Star Date, and the definition stuck. So when someone today talks about a blue moon, he or she is referring to the second full moon in a month.

?

Why are planets round?

ANSWERED BY:
Derek Sears, Professor of Cosmochemistry,
University of Arkansas, Fayetteville, Arkansas,
and Editor of Meteoritics and Planetary Science.

Planets are round because of their gravitational fields. A planet behaves like a fluid, and over long periods of time it succumbs to the gravitational pull from its center of gravity. The only way to get all the mass as close to the planet's center of gravity as possible is to form a sphere. The technical name for this process is "isostatic adjustment."

With much smaller bodies, such as the 20-kilometer asteroids we have seen in recent spacecraft images, the gravitational pull is too weak to overcome the asteroid's mechanical strength. As a result, these bodies do not form spheres, but they maintain irregular, fragmentary shapes.

?

How do scientists measure the weight of a planet?

ANSWERED BY:
**Gregory A. Lyzenga, Professor of Physics,
Harvey Mudd College, Claremont, California.**

The weight (or the mass) of a planet is determined by its gravitational effect on other bodies. Newton's Law of Gravitation states that every bit of matter in the universe attracts every other with a gravitational force that is proportional to its mass. For objects of the size we encounter in everyday life, this force is so minuscule that we don't notice it. However, for objects the size of planets or stars, it is of great importance.

In order to use gravity to find the mass of a planet, we must somehow measure the strength of its "tug" on another object. If the planet in question has a moon (a natural satellite), then nature has already done the work for us. By observing the time it takes for the satellite to orbit its primary planet, we can utilize Newton's equations to infer what the mass of the planet must be.

For planets without observable natural satellites, we must be more clever. Although Mercury and Venus, for example, do not have moons, they do exert a small pull on each other and on the other planets in the solar system. As a result, the planets follow paths that are subtly different than they would be without this disturbing effect. Although the mathematics is a bit more difficult, and the uncertainties are greater, astronomers can use these small deviations to determine how massive the moonless planets are.

What about those objects such as asteroids, whose masses are so small that they do not measurably disturb the orbits of the other planets? Until recent years, the masses of such objects were simply estimates, based upon the apparent diameters and assumptions about the possible mineral makeup of those bodies.

Now, however, several asteroids have been (or soon will be) visited by spacecraft. Just like a natural moon, a spacecraft flying by an asteroid has its path bent by an amount controlled by the mass of the asteroid. This "bending" is measured by careful tracking and Doppler radio measurement from the earth.

?

How fast is the earth moving?

ANSWERED BY:
Rhett Herman, Professor of Physics, Radford University, Radford, Virginia.

Questions about how fast the earth—or anything, for that matter—is moving are incomplete unless they also ask, "Compared to what?" Without a frame of reference, questions about motion cannot be completely answered.

Consider the movement of the earth's surface with respect to the planet's center. The earth rotates once every 23 hours, 56 minutes, and 4.09053 seconds, and its circumference is roughly 40,075 kilometers. Thus, the surface of the earth at the equator moves at a speed of 460 meters per second—or roughly 1,000 miles per hour.

As schoolchildren, we learn that the earth is moving about our sun in a very nearly circular orbit. It covers this route at a speed of nearly 30 kilometers per second, or

67,000 miles per hour. In addition, our solar system—
Earth and all—whirls around the center of our galaxy at
some 220 kilometers per second, or 490,000 miles per
hour. As we consider increasingly large size scales, the
speeds involved become absolutely huge!

The galaxies in our neighborhood are also rushing at a
speed of nearly 1,000 kilometers per second toward a
structure called the Great Attractor, a region of space
roughly 150 million light-years (one light-year is about six
trillion miles) away from us. This Great Attractor, having
a mass 100 quadrillion times greater than our sun and a
span of 500 million light-years, is made of both the visible
matter that we can see and the so-called dark matter that
we cannot see.

Each of the motions described above were given relative to some structure. Our motion about our sun was
described relative to our sun, while the motion of our local
group of galaxies was described as toward the Great
Attractor. The question arises: Is there some universal
frame of reference relative to which we can define the
motions of all things? The answer may have been provided
by the Cosmic Background Explorer (COBE) satellite.

In 1989, the COBE satellite was placed in orbit about
the earth (again, the earth is the frame of reference!) to measure the long-diluted radiation echo of the birth of our universe. This radiation, which remains from the immensely
hot and dense primordial fireball that was our early universe, is known as the cosmic microwave background radia-

tion (CBR). The CBR presently pervades all of space. It is the equivalent of the entire universe "glowing with heat."

One of COBE's discoveries was that the earth was moving with respect to this CBR with a well-defined speed and direction. Because the CBR permeates all space, we can finally answer the original question fully, using the CBR as the frame of reference.

The earth is moving with respect to the CBR at a speed of 390 kilometers per second. We can also specify the direction relative to the CBR. It is more fun, though, to look up into the night sky and find the constellation known as Leo (the Lion). The earth is moving toward Leo at the dizzying speed of 390 kilometers per second. It is fortunate that we won't hit anything out there during any of our lifetimes!

?

Why and how do planets rotate?

ANSWERED BY:
George Spagna, Chair, Department of Physics,
Randolph-Macon College, Ashland, Virginia.

S tars and planets form in the collapse of huge clouds of interstellar gas and dust. The material in these clouds is in constant motion, and the clouds themselves are in motion, orbiting in the aggregate gravity of the galaxy. As a result of this movement, the cloud will most likely have some slight rotation as seen from a point near its center. This rotation can be described as "angular momentum."

As an interstellar cloud collapses, it fragments into smaller pieces, each collapsing independently and each carrying part of the original angular momentum. The rotating clouds flatten into protostellar disks, out of which individual stars and their planets form. By a mechanism not fully understood, but believed to be associated with the strong magnetic fields associated with a young star, most of the angular momentum is transferred into the remnant disks. Planets form from material in this disk, through accretion of smaller particles.

In our solar system, the giant gas planets (Jupiter, Saturn, Uranus, and Neptune) spin more rapidly on their axes than the inner planets do and possess most of the system's angular momentum. The sun itself rotates slowly, only once a month. The planets all revolve around the sun in the same direction and in virtually the same plane. They also all rotate in the same general direction, with the exceptions of Venus and Uranus. These differences are believed to stem from collisions that occurred late in the planets' formation.

Star Light, Star Bright
stars

?

What exactly is the North Star?

ANSWERED BY:

Rich Schuler, Adjunct Instructor and Outreach
Coordinator, Department of Physics and Astronomy,
University of Missouri-St. Louis.

The North Star, or Polaris, is the brightest star in the constellation Ursa Minor, the Little Bear (also known as the Little Dipper). As viewed by observers in the Northern Hemisphere, Polaris occupies a special place. It lies roughly one-half degree from the North Celestial Pole (NCP), the point in the night sky directly in line with the projection of the earth's axis. As the earth rotates on its axis (once every 24 hours), the stars in the northern sky appear to revolve around the NCP, so this particular star appears to remain stationary hour after hour and night after night.

Because the earth is spherical, the position of Polaris relative to the horizon depends on the location of an observer. For example, when viewed from the equator

(0 degrees latitude), Polaris lies on the northern horizon. As the observer moves northward from the equator—say, to Houston, Texas (30 degrees latitude)—Polaris is located 30 degrees above the northern horizon. This trend continues until the traveler reaches the geographic (not magnetic) North Pole. At this point, Polaris is 90 degrees above the northern horizon and appears directly overhead.

A traveler on land or sea need only measure the angle between the northern horizon and Polaris to determine his or her latitude. Thus, Polaris is a handy tool for finding the northern extent of one's position, or latitude, and was therefore heavily utilized by travelers in the past—especially sailors.

There is currently no known star in the Southern Hemisphere that coincides with the South Celestial Pole. Also, Polaris is not an absolute guide to measuring latitude on the earth for Northern Hemisphere observers. This is because the axis of the earth precesses in a conical motion. The location of the North (and South) Celestial Pole is defined by projecting the axis of the earth onto the celestial sphere; consequently, as the axis changes position, so, too, does the "North Star." As a result, 5,000 years ago the earth's axis pointed toward the star Draco, and the star Thuban was the North Star. Similarly, in 12,000 years the star Vega (in the constellation Lyra) will be the North Star.

?

How long do stars usually live?

ANSWERED BY:

John Graham, Astronomer, Carnegie Institution of Washington, Washington, D.C.

The length of a star's life depends on how fast it uses up its nuclear fuel. Our sun, in many ways an average sort of star, has been around for nearly five billion years and has enough fuel to keep going for another five billion years. Almost all stars shine as a result of the nuclear fusion of hydrogen into helium. This takes place within their hot, dense cores where temperatures are as high as 20 million degrees. The rate of energy generation for a star is very sensitive to both temperature and the gravitational compression from its outer layers. These parameters are higher for heavier stars, and the rate of energy genera-tion—and in turn the observed luminosity—goes roughly as the cube of the stellar mass. Heavier stars thus burn their fuel much faster than less massive ones do and are disproportionately brighter. Some will exhaust their avail-able hydrogen within a few million years. On the other hand, the least massive stars that we know are so parsimo-nious in their fuel consumption that they can live to ages older than that of the universe itself—about 15 billion years. But because they have such low energy output, they are very faint.

When we look up at the stars at night, almost all of the ones we can see are intrinsically more massive and brighter than our sun. Most longer-lasting stars that are fainter than the sun are just too dim to view without a telescope. At the end of a star's life, when the supply of available hydrogen is nearly exhausted, it swells up and brightens. Many stars that are visible to the naked eye are in this stage of their life cycles because this bias brings them preferentially to our attention. They are, on average, a few hundred million years old and slowly coming to the end of their lives. A massive star such as the red Betelgeuse in Orion, in contrast, approaches its demise much more quickly. It has been spending its fuel so extravagantly that it cannot be older than about 10 million years. Within a million years, it is expected to go into complete collapse before probably exploding as a supernova.

Stars are still being born at the present time from dense clouds of dust and gas, but they remain deeply embedded in their placental material and cannot be seen in visible light. The enveloping dust is transparent to infrared radiation, however, so scientists using modern detecting devices can easily locate and study them. In so doing, we hope to learn how planetary systems like our own come together.

?

Why do stars twinkle?

ANSWERED BY:
**John Graham, Astronomer, Carnegie Institution
of Washington, Washington, D.C.**

Have you ever noticed how a coin at the bottom of a swimming pool seems to wobble? This occurs because the water in the pool bends the path of light reflected from the coin. Similarly, stars twinkle because their light has to pass through several miles of Earth's atmosphere before it reaches the eye of an observer. It is as if we are looking at the universe from the bottom of a swimming pool. Our atmosphere is turbulent, with streams and eddies forming, churning, and dispersing all the time. These disturbances act like lenses and prisms that shift a star's light from side to side by minute amounts several times a second. For large objects such as the moon, these deviations average out. (Through a telescope with high magnification, however, the objects appear to shimmer.) Stars, in contrast, are so far away that they effectively act as point sources, and the light we see flickers in intensity as the incoming beams bend rapidly from side to side. Planets such as Mars, Venus, and Jupiter, which

appear to us as bright stars, are much closer to Earth and
look like measurable disks through a telescope. Again, the
twinkling from adjacent areas of the disk averages out,
and we see little variation in the total light emanating from
the planet.

Far, Far Away . . .
the universe

?

How do we know our location within
the Milky Way galaxy?

ANSWERED BY:
Laurence A. Marschall, Department of Physics,
Gettysburg College, Gettysburg, Pennsylvania.

Finding one's location in a cloud of a hundred billion
stars—when one can't travel beyond one's own
planet—is like trying to map out the shape of a forest
while tied to one of the trees. One gets a rough idea of the
shape of the Milky Way galaxy by just looking around—a
ragged, hazy band of light circles the sky. It is about 15

degrees wide, and stars are concentrated fairly evenly along the strip. That observation indicates that our Milky Way galaxy is a flattened disk of stars, with us located somewhere near the plane of the disk. Were it not a flattened disk, it would look different. For instance, if it were a sphere of stars, we would see its glow all over the sky, not just in a narrow band. And if we were above or below the disk plane by a substantial amount, we would not see it split the sky in half—the glow of the Milky Way would be brighter on one side of the sky than on the other.

The position of the sun in the Milky Way can be further pinned down by measuring the distance to all the stars we can see. In the late eighteenth century, the astronomer William Herschel tried to do this, concluding that the earth was in the center of a "grindstone"-shaped cloud of stars. But Herschel was not aware of the presence of small particles of interstellar dust, which obscure the light from the most distant stars in the Milky Way. We appeared to be in the center of the cloud because we could see no further in all directions. To a person tied to a tree in a foggy forest, it looks like the forest stretches equally away in all directions, wherever one is.

A major breakthrough in moving the earth from the center of the galaxy to a point about three-fifths away from the edge came in the early decades of the twentieth century, when the astronomer Harlow Shapley measured the distance to the large clusters of stars called globular clusters. He found that they were distributed in a spherical distribution about

100,000 light-years in diameter, centered on a location in the constellation Sagittarius. Shapley concluded (and other astronomers have since verified) that the center of the distribution of globular clusters is the center of the Milky Way as well, so our galaxy looks like a flat disk of stars embedded in a spherical cloud, or "halo," of globular clusters.

In the past 75 years, astronomers have refined this picture, using a variety of techniques of radio, optical, infrared, and even x-ray astronomy, to fill in the details: the location of spiral arms, clouds of gas and dust, concentrations of molecules, and so on. The essential modern picture is that our solar system is located on the inner edge of a spiral arm, about 25,000 light-years from the center of the galaxy, which is in the direction of the constellation of Sagittarius.

?

Why is the night sky dark?

ANSWERED BY:
Karen B. Kwitter, Ebenezer Fitch
Professor of Astronomy, Williams College,
Williamstown, Massachusetts.

We see stars all around, so why doesn't their combined light add up to make our night sky—and surrounding space, for that matter—bright? The German physicist Heinrich Wilhelm Olbers put the same puzzle this way in 1823: If the universe is infinite in size, and stars (or galaxies) are distributed throughout this infinite universe, then we are certain to eventually see a star in any direction we look. As a result, the night sky should be aglow. Why isn't it?

In fact, the answer is far more profound than it appears. There have been many attempts at explaining this puzzle, dubbed Olbers' Paradox, over the years. One version implicated dust between stars and perhaps between galaxies. The idea was that the dust would block the light from faraway objects, making the sky dark. In reality, however, the light falling on the dust would eventually heat it up so that it would glow as brightly as the original sources of the light.

Another answer proposed that the tremendous red shift of distant galaxies—the lengthening of the wavelength of light they emit due to the expansion of the universe—would move light out of the visible range into the invisible infrared. But if this explanation were true, shorter-wavelength ultraviolet light would also be shifted into the visible range—which doesn't happen.

The best resolution to Olbers' Paradox at present has two parts. First, even if our universe is infinitely large, it is not infinitely old. This point is critical because light travels at the finite (though very fast!) speed of about 300,000 kilometers per second. We can see something only after the light it emits has had time to reach us. In our everyday experience that time delay is minuscule: even seated in the balcony of the concert hall, you will see the conductor of the symphony raise her baton less than a millionth of a second after she actually does.

When distances increase, though, so does the time delay. For instance, astronauts on the moon experience a 1.5-second time delay in their communications with Mission Control due to the time it takes the radio signals (which are a form of light) to travel round-trip between earth and the moon. Most astronomers agree that the universe is between 10 and 15 billion years old. And that means that the maximum distance from which we can receive light is between 10 and 15 billion light-years away. So even if there are more distant galaxies, their light will not yet have had time to reach us.

The second part of the answer lies in the fact that stars and galaxies are not infinitely long-lived. Eventually, they will dim. We will see this effect sooner in nearby galaxies, thanks to the shorter light-travel time. The sum of these effects is that at no time are all of the conditions for creating a bright sky fulfilled. We can never see light from stars or galaxies at all distances at once; either the light from the most distant objects hasn't reached us yet, or if it has, then so much time would have had to pass that nearby objects would be burned out and dark.

?

Does the fact that the universe is continually expanding mean that it lacks a physical edge?

ANSWERED BY:
Stephen Reucroft and John Swain,
Professors of Physics, Northeastern University,
Boston, Massachusetts.

A tricky part of this question is the wording: "The universe," by definition, is all there is! To speak of the universe expanding into something would mean that there was something bigger, which we ought to have called "the universe" in the first place.

Perhaps the easiest way to see what is meant by an expanding universe is to imagine what life would be like for two-dimensional ants living on the surface of an expanding spherical balloon. They can crawl around, but being unable to fly, or to penetrate the balloon's surface, they live in what is essentially a two-dimensional world. For the ants, provided nothing disturbs them from outside, the universe is the surface of the balloon—that's all there is! Being confined to the surface of the balloon, there is no way for the ants to discover anything at all about what we would term up and down.

The area of this two-dimensional universe is finite,

but nowhere will the ants find a boundary or an edge. Here you have to ignore the rubber neck of the balloon and the person blowing it up; think of a balloon sealed smoothly into a spherical shape hovering in a tank in which the air pressure could be lowered to make the balloon expand.

Now as the balloon expands, the ants see one another getting farther and farther apart. Each sees the same thing: All its neighbors are moving away. The ants live in what for them is an expanding universe with no physical edge. If an ant walks quickly enough, it could conceivably get all the way around the balloon and return to its starting point without encountering an edge.

You may object at this point and claim that we see the balloon expanding into the surrounding space. But we have access to an extra dimension in which to move—the one that would correspond to up and down for the ants, were they able to move in those directions. As far as the ants are concerned, they can learn everything they want to know about their world by making measurements on the balloon's surface, with no reference to the surrounding space.

Experience in physics has taught us that when we find a concept that is spurious—in the sense that it leads to no predictable effects—we do better just to assume it's not there. In other words, the ants would do well not to talk about their space expanding into something that they can't measure. Nothing is lost and there is a substantial gain in simplicity.

To make an analogy between the ant's situation and our own, you have to imagine space expanding in all directions. Everybody in the universe sees everything rushing away from everything else; but the universe need have no physical edges, and there is no need to describe it as expanding into anything; it can just expand.

2

It's Alive!

The Grass Is Always Greener
plants

?

What causes the leaves on trees to change color in the fall?

ANSWERED BY:
**Alan Dickman, Curriculum Director, Department of
Biology, University of Oregon, Eugene.**

Leaves of all trees contain chlorophyll, a green pigment that has the unusual capability to capture light energy and—with the help of other components in the leaf—to convert that energy into a chemical form, such as sugar. Many leaves contain other pigments as well, and while these pigments can't photosynthesize as chlorophyll can, some of them are able to transfer the light energy they capture to the chlorophyll. Some of these "accessory" pigments are yellow, orange, or red and are called carotenoids because they belong to the same group of compounds as beta-carotene, the pigment that gives carrots their orange color.

In the autumn, when leaves begin to get old, the leaf is able to break down some of the extensive pigments it has produced (such as chlorophyll) and absorb parts of them back into the stems for other uses. When the green color of chlorophyll is gone, the other colors are unmasked.

You can see these colors when the leaves are still green if you separate the pigments by a process called chromatography. If you have ever watched water-soluble ink smear on paper when it gets wet, you have seen chromatography in action. Separating the pigments from leaves is a little harder, because they are often enclosed in membranes within the cells of a leaf. But if you have some filter paper (try using a white coffee filter) you could try to express some of the pigments onto it by placing the leaf on the filter and then rolling a quarter across the leaf several times to make a line of pigments on the paper. Then dip one end of the paper in rubbing alcohol, and you might be able to see some of the other colors in the leaf separate from the green chlorophyll.

Some pigments in leaves—such as the reddish-purple in rhubarb or red cabbage—are not involved in photosynthesis at all. Their purpose is unclear, but it's likely they help protect the plant against too much sunlight. These compounds are held in other places in the cells of the leaf, and many of them are water-soluble, so if you cook the leaf or grind it in a blender, you will release this reddish pigment in the water.

?

How does the Venus flytrap digest flies?

ANSWERED BY:
Lissa Leege, Plant Ecologist and Assistant Professor of Biology, Georgia Southern University, Statesboro.

Perhaps the best known of the insectivorous (insect-eating) plants, the Venus flytrap (*Dionaea muscipula*) exhibits a unique system by which it attracts, kills, digests, and absorbs its prey. Because it is a plant and can make its own food through photosynthesis, the Venus flytrap does not eat and digest its prey for the traditional nonplant objectives of harvesting energy and carbon. Instead, it mines its prey primarily for essential nutrients (nitrogen and phosphorous in particular) that are in short supply in its boggy, acidic habitat. So, yes, the Venus flytrap does have a digestive system of sorts, but it serves a somewhat different purpose than an animal's does.

How does a stationary organism manage to attract, kill, digest, and absorb its prey? First, it lures its victim with sweet smelling nectar, secreted on its steep-trap-shaped leaves. Unsuspecting prey land on the leaf in search of a reward but instead trip the bristly trigger hairs on the leaf and find themselves imprisoned behind the interlocking teeth of the leaf edges. There are between three and six trigger hairs on the surface of each leaf. If the same hair is touched twice, or if two hairs are touched within a 20-second interval, the cells on the outer surface of the leaf expand rapidly, and the trap snaps shut instantly. If insect secretions stimulate the trap, it will clamp down further on the prey and form an airtight seal. (If tripped by a curious spectator or a falling dead twig, the trap will reopen within a day or so.)

Once the trap closes, the digestive glands that line the interior edge of the leaf secrete fluids that dissolve the soft parts of the prey, kill bacteria and fungi, and break down the insect with enzymes to extract the essential nutrients. These nutrients are absorbed into the leaf, and five to twelve days following capture, the trap will reopen to release the leftover exoskeleton. After three to five meals, the trap will no longer capture prey but will spend another two to three months simply photosynthesizing before it drops off the plant.

?

How do trees carry water from the soil around their roots to the leaves at the top?

ANSWERED BY:
Alan Dickman, Curriculum Director, Department of Biology, University of Oregon, Eugene.

Once inside the cells of the root, water enters into a system of interconnected cells that make up the wood of the tree and extend from the roots through the stem and branches and into the leaves. The scientific name for wood tissue is xylem; it consists of a few different kinds of cells. The cells that conduct water (along with dissolved mineral nutrients) are long and narrow and are no longer alive when they function in water transport. Some of them have open holes at their tops and bottoms and are stacked more or less like concrete sewer pipes. Other cells taper at their ends and have no complete holes. All have pits in their cell walls, however, through which water can pass. Water moves from one cell to the next when there is a pressure difference between the two.

Because these cells are dead, they cannot be actively involved in pumping water. It might seem possible that living cells in the roots could generate high pressure in the root cells, and to a limited extent this process does occur. But common experience tells us that water within the wood is not under positive pressure—in fact, it is under negative pressure, or suction. To convince yourself of this, consider what happens when a tree is cut or when a hole is drilled into the stem. If there were positive pressure in the stem, you would expect a stream of water to come out, which rarely happens.

In reality, the suction that exists within the water-conducting cells arises from the evaporation of water molecules from the leaves. Each water molecule has both positive and negative electrically charged parts. As a result, water molecules tend to stick to one another; that adhesion is why water forms rounded droplets on a smooth surface and does not spread out into a completely flat film. As one water molecule evaporates through a pore in a leaf, it exerts a small pull on adjacent water molecules, reducing the pressure in the water-conducting cells of the leaf and drawing water from adjacent cells. This chain of water molecules extends all the way from the leaves down to the roots and even extends out from the roots into the soil. So the simple answer is that the sun's energy does it: Heat from the sun causes the water to evaporate, setting the water chain in motion.

Creepy Crawlers
insects

?

How is bug blood different from our own?

ANSWERED BY:
Rob DeSalle, Curator, Division of Invertebrate Zoology, American Museum of Natural History, New York City.

The major difference between insect blood and the blood of vertebrates, including humans, is that vertebrate blood contains red blood cells. Insects and other invertebrates, on the other hand, have what is called hemolymph—a heterogeneous fluid that courses through their bodies, bathing all the internal tissues. Hemolymph is mostly water, but it also contains ions, carbohydrates, lipids, glycerol, amino acids, hormones, some cells, and pigments. The pigments, however, are usually rather bland, and thus insect blood is clear or tinged with yellow or green. (The red color that you see upon squashing a housefly or fruit fly is actually pigment from the insect's eyes.)

Unlike the closed circulatory system found in vertebrates, insects have an open system lacking arteries and veins. The hemolymph thus flows freely throughout their bodies, lubricating tissues and transporting nutrients and wastes. Whereas the vertebrate circulatory system serves primarily to carry oxygen throughout the body, insects respire an entirely different way—namely, through tracheal tubes. In the case of the fruit fly *(Drosophila)*, for example, a series of tiny openings called spiracles line the impermeable outer skin of the fly, and these convey air directly to tracheal tubes, which, in turn, convey air to the internal tissues.

Insects do have hearts that pump the hemolymph throughout their circulatory systems. Though these hearts are quite different from vertebrate hearts, some of the genes that direct heart development in the two groups are in fact very similar. The development and evolution of the vertebrate heart is currently the subject of much research.

?

What kind of illnesses do insects get?

ANSWERED BY:

Deborah A. Kimbrell, Assistant Professor of Biology, University of Houston.

Bacteria and viruses can be problems for insects; some insects also face threats from parasitic wasps or other parasites. Insects, however, have very effective immune systems for fighting illness. When bacteria enters through a wound, insect blood cells are quickly mobilized to surround and digest the invading bacteria. At the same time, the fat body—a tissue analogous to the mammalian liver—produces large quantities of antibacterial proteins. This immune response of insects is very similar to that in mammals, and in recent years scientists have studied the common fruitfly *(Drosophila melanogaster)* to learn about the fundamentals of immunity in both insects and mammals.

?

How do flies and other insects walk up walls?

ANSWERED BY:
Richard D. Fell, Associate Professor,
Department of Entomology, Virginia Polytechnic and
State University, Blacksburg.

Numerous insects, such as common houseflies, as well as certain amphibians and reptiles, are able to walk on and cling to seemingly smooth surfaces—including glass doors and windows. The segments, or tarsi, at the end of insect legs feature clawlike structures that help the insect hold on to different types of surfaces. These tarsal claws are used to grip the tiny irregularities on rough surfaces. But in some cases, insects do make use of a kind of adhesion. If the surface is smooth, the insect can hold on using the adhesive action of hairs located on sticky pads (known as the arolia or pulvilli) on the tarsi.

Some insects, such as grasshoppers, have pads on each of their tarsal segments, and some insects may have special adhesive pads on other segments of the leg. The pads typically contain numerous hairs that secrete an oily substance that causes the tips of the hairs to adhere to the surface. This substance provides the traction and stickiness that allows insects to hold on to smooth surfaces, such as glass. Surfaces that appear perfectly smooth to us actually have many microscopic bumps and fissures, which serve as footholds for the tiny hairs.

?

Why is spider silk so strong?

ANSWERED BY:
William K. Purves, Biologist, Harvey Mudd College, Claremont, California.

Dragline silk, the silk that forms the radial spokes of a spider's web, is composed of two proteins, making it strong and tough—yet elastic. Each protein contains three regions with distinct properties. The first forms an amorphous (noncrystalline) matrix that is stretchable, giving the silk elasticity. When an insect strikes the web, the matrix stretches, absorbing the kinetic energy proteins. There are two kinds of crystalline regions that toughen the silk. Both regions are tightly pleated and resist stretching, and one of them is rigid. It is thought that the pleats of the less rigid crystalline regions anchor the rigid crystals to the matrix.

A spider's dragline is only about one tenth the diameter of a human hair, but it is several times stronger than steel, on a weight-for-weight basis. The movie *Spider-Man* drastically underestimates the strength of silk—real dragline silk would not need to be nearly as thick as the strands deployed by our web-swinging hero.

?

If a used needle can transmit HIV, why can't a mosquito?

ANSWERED BY:

Laurence Corash, Chief Medical Officer, Cerus Corporation, Concord, California.

The AIDS virus (HIV) on used needles is infectious when injected into a human where the virus can bind to T cells and start to replicate. The human T cell is a very specific host cell for HIV. When a mosquito feeds on a person with HIV in his or her blood, the HIV enters the insect's gut, which does not contain human T cells. The virus thus has no host cell in which to replicate and it is broken down by the mosquito's digestive system.

The single-celled parasite that causes malaria, in contrast, can survive and multiply in the mosquito's gut and mature into an infectious form. The resulting sporozoites then migrate to the insect's salivary glands. Because mosquitoes inject their saliva when they bite, the parasite is passed along to the next human the insect feeds on. In this case the complex interaction between the infectious agent and the vector (the mosquito) is required for transmission. HIV, however, deteriorates in the gut before the mosquito bites again and therefore is not transmitted to the insect's next victim.

Under the Sea
ocean life

?

How do squid and octopuses change color?

ANSWERED BY:
Ellen J. Prager, Assistant Dean, Rosensteil School of Marine and Atmospheric Science, University of Miami, and Author of *The Oceans*.

A number of cephalopods—the group of animals that includes octopuses, squid, and cuttlefish—are skilled in the art of color change, which can be used for camouflage or to startle and warn potential predators. Many of these creatures have special pigment cells called chromatophores in their skin. By controlling the size of the cells they can vary their color and even create changing patterns. Chromatophores are connected to the nervous system, and their size is determined by muscular contractions. The cephalopods also have extremely well-developed eyes, which are believed to detect both the color and intensity of light. Using their excellent eyesight and chromatophores, cephalopods camouflage themselves by creating color patterns that closely match the underlying seafloor. In squid, color changes also occur when the animal is disturbed or feels threatened.

In addition to color change, many squid can produce light and control its intensity. Biologically produced light is called bioluminescence, and it is used for a wide variety of purposes by marine organisms. Some creatures are believed to use bioluminescence to confuse or startle predators, others may stun their prey, and some use it as a decoy to facilitate escape or as a lure to attract the unwary. Bioluminescence may also offer a means of communication in the dim midwater or twilight region of the sea.

Squid and other marine creatures create light by mixing two substances into a third that gives off light, similar to the mechanism by which a common firefly lights up or the way the popular plastic green glowsticks work. To get a glowstick to "glow," it is bent. This causes the two chemicals inside to mix and react, yielding a third substance that gives off light. Within an organism's special light-producing cells (photocytes) or organs (photophores), essentially the same thing happens. A substance called luciferin reacts with oxygen in the presence of an enzyme, luciferase. A new molecule forms when the reaction is complete, and in the ocean it typically glows blue to green in color. In some organisms the photophores are simple glandular cups. In others they are elaborate devices with lenses for focusing, a color filter, or an adjustable flap that serves as an off/on switch. Squid that have both photophores and chromatophores within their skin can control both the color and the intensity of light produced. Research has also revealed that within some squid and fish, bioluminescent light may be produced by bacteria that live inside the animal's light organs.

?

Why do some fish normally live in freshwater and others in saltwater? How can some fish adapt to both?

A N S W E R E D   B Y :

Aldo Palmisano, Research Chemist, Western Fisheries Research Center, U.S. Geological Survey Biological Resources Division. Palmisano is affiliated with the University of Washington, Seattle.

The reason some fish normally live in freshwater and others live in seawater is that one or the other environment provides them with opportunities that have traditionally contributed to their survival. An obvious difference between the two habitats is salt concentration. Freshwater fish maintain the physiological mechanisms that permit them to concentrate salts within their bodies in a salt-deficient environment; marine fish, on the other hand, excrete excess salts. Fish that live in both environments retain both mechanisms.

Life began evolving several billion years ago in the oceans and since that time, living things have maintained an internal environment closely resembling the ionic composition of those primeval seas. Presumably, the ionic conditions in which life began are uniquely appropriate to its continuation. Laboratory studies support the view that the various chemical phenomena on which life depends—including the interactions of nucleic acids with each other and with proteins, the folding and performance of proteins such as enzymes, the functioning of intracellular machines such as ribosomes, and the maintenance of cellular compartments—are critically dependent on the ionic milieu in which the reactions take place.

Given time, ocean-dwelling creatures took advantage of untapped resources, such as relatively safe spawning habitats or new food sources, that were available to them only by colonizing other environments, like freshwater and land. Colonization was facilitated, if not necessitated, by geological events, such as the movements and collisions of land masses (plate tectonics) and volcanic activity, which served to isolate portions of very similar populations of a single species from one another. Such geological change forced some populations to either adapt or face extinction. Time and natural selection due to physical and environmental variation worked in concert with isolation to foster adaptations. In some cases, these adaptations became permanent and led to species differentiation.

One important aspect of environmental variation is the ionic composition of bodies of water utilized as habitat. Chloride cells in the gills of marine fish produce an enzyme, called gill $Na+/K+$ ATPase, that enables them to rid their plasma of excess salt, which builds up when they drink seawater. They use the enzyme to pump sodium out of their gills at the cost of energy. Additionally, their kidneys selectively filter out divalent ions, which they then excrete. An alternative set of physiological mechanisms allows freshwater fish to concentrate salts to compensate for their low salinity environment. They produce very dilute, copious urine (up to a third of their body weight a day) to rid themselves of excess water, while conducting active uptake of ions at the gill.

Certainly, other adaptations contributed to the capability of isolated populations to adapt more fully to their circumstances. With different sets of predator and prey organisms present in the differing habitats, and different physical ranges available to them, behavioral changes would be required; perhaps a smaller or larger body size or body part would be favored. The accumulation of these kinds of physiological, behavioral, and physical changes ultimately led to new species. Isolation may have forced them to conserve their newly developed adaptations among their own descendants, rather than distribute them more broadly. For some, the rift eventually became complete and there could no longer be any cross-breeding between populations that once interbred.

Not unreasonably, there were multiple instances of colonization of the freshwater environment by seawater species of fish; some were more or less complete. The ability to escape an environment may have been seasonal, or periodic in some other way, or intermittent. The ability to osmoregulate in freshwater need not have excluded the capacity to revert to a seawater mode of osmoregulation, as long as the capacity could be utilized by a substantial portion of the population and selected for, rather than simply lost.

?

How can sea mammals drink saltwater?

ANSWERED BY:
Robert Kenney, Marine Biologist, University of Rhode Island, Kingston, Rhode Island.

Although some marine mammals are known to drink seawater at least on occasion, it is not well established that they routinely do so. They have other options: Sea-dwelling mammals can get water through their food, and they can produce it internally from the metabolic breakdown of food (water is one of the by-products of carbohydrate and fat metabolism).

The salt content of the blood and other body fluids of marine mammals is not very different from that of terrestrial mammals or any other vertebrates: It is about one-third as salty as seawater. Because a vertebrate that drinks seawater is imbibing something three times saltier than its blood, it must get rid of the excess salt by producing very salty urine. The urine of seals and sea lions, for instance, contains up to two and a half times more salt than seawater does and seven or eight times more salt than their blood.

Salt and water management in mammalian kidneys is a two-step process. First the blood passes through a micro-filter system in a part of the kidney known as the glomerulus. Most of the blood plasma, including water and small molecules like salts, passes through the filter, but the larger molecules, as well as the blood cells, are held back. The filtered plasma then passes through a long tube called the loop of Henle, where the water is reabsorbed. This process concentrates the remaining fluid, which is finally excreted as urine. One popular theory holds that a simple modification of the standard mammalian kidney—namely, longer loops of Henle—allows marine mammals to produce a more concentrated urine by reclaiming more of the water. Kidney anatomy in manatees and harbor porpoises seems to support this theory, but it has not been closely studied in most marine mammal species.

A marine mammal can minimize its salt and water balance problems by following the same advice doctors give patients to keep their blood pressure down: avoid salty food. With the exception of the herbivorous manatees and dugongs, all marine mammals are carnivores. Different food types vary in salt content. Species that subsist on plants or invertebrates (such as crustaceans and mollusks) consume food with about the same salt content as seawater. These species face the same salt removal problem they would have if they drank seawater directly. In contrast, marine mammals that feed on fish consume food with a salt content similar to that of their own blood, thereby avoiding the problem entirely. Indeed, a study of California sea lions showed that, on a diet of fish, these animals can live without drinking fresh water at all.

Some species of seals and sea lions apparently do drink seawater at least occasionally, as do common dolphins and sea otters, but the practice is very rare in some other species. When given the choice, manatees and some pinnipeds will drink fresh water. (People who live on salt or brackish waterways in Florida sometimes leave a garden hose flowing into the water in order to see the manatees come to drink.) Likewise, some seals will eat snow to get fresh water. For most whales and dolphins, however, we simply do not know how they get their water, because it is difficult to observe these animals.

?

How do deep-diving sea creatures withstand huge pressure changes?

ANSWERED BY:
Paul J. Ponganis and Gerald L. Kooyman, Center for Marine Biotechnology and Biomedicine, Scripps Institution of Oceanography, La Jolla, California.

The biggest challenges in adapting to pressure are probably faced by those animals that must routinely travel from the surface to great depth, such as the sperm whale and the bottlenose whale. From the days of whaling, these animals have been recognized as exceptional divers, with reports of dives lasting as long as two hours after they were harpooned. Today, with the use of sonar tracking and attached time-depth recorders, dives as deep as 6,000 feet (more than a mile below the surface of the ocean) have been measured. Routine dive depths are usually in the 1,500- to 3,000-foot range, and dives can last between 20 minutes and an hour.

The primary anatomical adaptations for pressure of a deep-diving mammal center on air-containing spaces and the prevention of tissue barotrauma. Air cavities, when present, are lined with venous plexuses, which are thought to fill at depth, obliterate the air space, and prevent "the squeeze." The lungs collapse, which prevents lung rupture and (important with regard to physiology) blocks gas exchange in the lung. Lack of nitrogen absorption at depth prevents the development of nitrogen narcosis and decompression sickness. In addition, because the lungs do not serve as a source of oxygen at depth, deep divers rely on enhanced oxygen stores in their blood and muscle.

How do whales and dolphins sleep without drowning?

ANSWERED BY:

Bruce Hecker, Director of Husbandry, South Carolina Aquarium, Charleston.

Marine mammals such as whales and dolphins spend their entire lives at sea. So how can they sleep and not drown? Observations of bottlenose dolphins in aquariums and zoos, and of whales and dolphins in the wild, show two basic methods of sleeping: They either rest quietly in the water, vertically or horizontally, or sleep while swimming slowly next to another animal. Individual dolphins also enter a deeper form of sleep, mostly at night. It is called logging because in this state, a dolphin resembles a log floating at the water's surface.

When marine mammals sleep and swim at once, they are in a state similar to napping. Young whales and dolphins actually rest, eat, and sleep while their mother swims, towing them along in her slipstream. At these times, the mother will also sleep on the move. In fact, she cannot stop swimming for the first several weeks of a newborn's life. If she does for any length of time, the calf will begin to sink; it is not born with enough body fat or blubber to float easily.

Lots of swimming will tire an infant, producing a weak animal susceptible to infection or attack. Adult male dolphins, which generally travel in pairs, often swim slowly side by side as they sleep. Females and young travel in larger pods. They may rest in the same general area, or companionable animals may pair for sleeping while swimming.

While sleeping, the bottlenose dolphin shuts down only half of its brain, along with the opposite eye. The other half of the brain stays awake at a low level of alertness. This attentive side is used to watch for predators, obstacles, and other animals. It also signals when to rise to the surface for a fresh breath of air. After approximately two hours, the animal will reverse this process, resting the active side of the brain and awaking the rested half. This pattern is often called catnapping.

Dolphins generally sleep at night but only for a couple

of hours at a time; they are often active late at night, possibly matching this alert period to feed on fish or squid, which then rise from the depths. Bottlenose dolphins spend an average of one-third of their day asleep. It is not clear whether they undergo dream sleep. Rapid eye movement (REM)—a characteristic of deep sleep—is hard to discern. But a pilot whale was noted as having six minutes of REM in a single night.

To avoid drowning during sleep, it is crucial that marine mammals retain control of their blowhole. The blowhole is a flap of skin that is thought to open and close under the voluntary control of the animal. Although still a matter of discussion, most researchers feel that in order to breathe, a dolphin or whale must be conscious and alert to recognize that its blowhole is at the surface. Humans, of course, can breathe while the conscious mind is asleep; our subconscious mechanisms have control of this involuntary system. But equipped with a voluntary respiratory system, whales and dolphins must keep part of the brain alert to trigger each breath.

Other methods help marine mammals to hold their breath longer than other types of mammals can. Marine mammals can take in more air with each breath, because their lungs are proportionately larger than those in humans. In addition, they exchange more air with each inhalation and exhalation. Their red blood cells also carry more oxygen. And when diving, marine mammals' blood travels only to the parts of the body that need oxygen—the heart, the brain, and the swimming muscles. Digestion and any other processes have to wait.

Finally, these animals have a higher tolerance for carbon dioxide (CO_2). Their brains do not trigger a breathing response until the levels of CO_2 are much higher than what humans can tolerate. These mechanisms, part of the marine mammal diving response, are adaptations to living in an aquatic environment and help during the process of sleeping. They reduce the number of breaths taken during rest periods; a dolphin might average 8 to 12 breaths a minute when fairly active only to have their breathing rate drop to 3 to 7 per minute while resting.

It is actually rare for a marine mammal to drown, as they won't inhale underwater; but it is possible for them to suffocate from a lack of air. Being born underwater can cause problems for newborn whale and dolphin calves. It is the touch of air on the skin that triggers that first, crucial breath, and necropsies sometimes show that an animal never gets to the surface to take its first breath of air. The same problem can occur when an animal is caught in a fishing net. If unable to reach the surface, or if in a panic, the animal may dive deeper, where it will be unable to breathe and will suffocate.

That's a Horse of a Different Color
animal kingdom

?

Do hippopotamuses actually have pink sweat?

ANSWERED BY:
Mark Ritchie, Professor of Biology, Syracuse University, Syracuse, New York.

Hippos secrete a reddish oily fluid sometimes called "blood sweat" from special glands in their skin. But the fluid is not sweat. Unlike sweat, which some mammals (including humans) secrete onto their skin, where it evaporates and cools the body, this fluid functions as a skin moisturizer, a water repellent, and an antibiotic. It appears red when exposed to full sunlight, which led the first European discoverers in Africa to call it "blood sweat."

Hippos try to avoid direct sunlight by lying in water during the day and feeding at night. Their skin is very sensitive to both drying and sunburn, so the secretion acts like an automatic skin ointment. It also protects the skin from becoming waterlogged when a hippo is in the water. The detailed chemical composition of this secretion, which is unique to hippos, remains something of a mystery.

?

Why do cats purr?

ANSWERED BY:
Leslie A. Lyons, Assistant Professor, School of Veterinary Medicine, University of California, Davis.

Over the course of evolution, purring has probably offered some selective advantage to cats. Most feline species produce a "purr-like" vocalization. In domestic cats, purring is most noticeable when an animal is nursing her kittens or when humans provide social contact via petting, stroking, or feeding.

Although we assume that a cat's purr is an expression of pleasure or is a means of communication with its young, perhaps the reasons for purring can be deciphered from the more stressful moments in a cat's life. Cats often purr while under duress, such as during a visit to the veterinarian or when recovering from injury. Thus, not all purring cats appear to be content or pleased with their circumstances. This riddle has lead researchers to investigate how cats purr, which is also still under debate.

Scientists have demonstrated that cats produce the purr through intermittent signaling of the laryngeal and diaphragmatic muscles. Cats purr during both inhalation and exhalation with a consistent pattern and frequency between 25 and 150 hertz. Various investigators have shown that sound frequencies in this range can improve bone density and promote healing.

This association between the frequencies of cats' purrs and improved healing of bones and muscles may provide help for some humans. Bone density loss and muscle atrophy is a serious concern for astronauts during extended periods at zero gravity. Their musculoskeletal systems do not experience the normal stresses of physical activity, including routine standing or sitting, which requires strength for posture.

Because cats have adapted to conserve energy via long periods of rest and sleep, it is possible that purring is a low-energy mechanism that stimulates muscles and bones without a lot of energy. The durability of the cat has facilitated the notion that cats have "nine lives," and a common veterinary legend holds that cats are able to reassemble their bones when placed in the same room with all their parts. Purring may provide a basis for this feline mythology. The domestication and breeding of fancy cats occurred relatively recently compared to other pets and domesticated species; thus cats do not display as many muscle and bone abnormalities as their more strongly selected carnivore relative, the domestic dog. Perhaps cats' purring helps alleviate the dysplasia or osteoporotic conditions that are more common in their canid cousins.

?

Why do dogs get blue, not red, eyes in flash photos?

ANSWERED BY:

J. Phillip Pickett, Veterinary Ophthalmologist, Virginia-Maryland Regional College of Veterinary Medicine, Blacksburg, Virginia.

R ed eye," the all-too-familiar nemesis of amateur photographers, occurs when a person looks directly at the camera when his or her picture is taken. If the flash is on the same axis as the visual axis of the camera, the reflection of the light off the blood vessels in the person's retina will give an eerie, satanic "red eye" look. People with light-colored eyes usually exhibit the worst red eye effect; those individuals with dark-colored eyes may have enough pigment in the back of their eyes to mask this so-called red reflex.

Dogs, cats, and almost all domestic animals have a special reflective layer in the back of the eye termed the tapetum, which enhances nocturnal vision. Light passes through the animal's retina from outside of the eye and is then reflected back through the retina a second time from the reflective tapetal layer beneath the retina. This double stimulation of the retina helps these species to see better than humans do in dim light situations. The color of this tapetal layer varies to some extent with an animal's coat color. A black Labrador retriever, for example, will usually have a green tapetal reflection. A buff cocker spaniel will generally show a yellow tapetal reflection. Most young puppies and kittens have a blue tapetal reflection until the structures in the back of the eye fully mature at six to eight months of age. "Color dilute" dogs and cats, such as red Siberian huskies and blue point Siamese cats, may have no tapetal pigment and may therefore exhibit a red reflex just like human beings.

?

How do frogs survive winter? Why don't they
freeze to death?

A N S W E R E D   B Y :
Rick Emmer, Lead Keeper, The RainForest,
Cleveland Metroparks Zoo.

D espite their fragile appearance and inoffensive ways, frogs have countless strategies to deal with the most severe climates this planet has to offer. They can be found at the Arctic Circle, in deserts, in tropical rain forests, and practically everywhere in between. Some of their survival strategies are nothing short of ingenious. Various frog species use two strategies to deal with environmental extremes: hibernation and estivation.

Hibernation is a common response to the cold winter of temperate climates. After an animal finds or makes a living space (hibernaculum) that protects it from winter weather and predators, the animal's metabolism slows dramatically, so it can "sleep away" the winter by utilizing its body's energy stores. When spring weather arrives, the animal "wakes up" and leaves its hibernaculum to get on with the business of feeding and breeding.

Aquatic frogs such as the leopard frog and American

bullfrog typically hibernate underwater. A common mis-conception is that they spend the winter the way aquatic turtles do, dug into the mud at the bottom of a pond or stream. In fact, hibernating frogs would suffocate if they dug into the mud for an extended period of time. A hiber-nating turtle's metabolism slows down so drastically that it can get by on the mud's meager oxygen supply. Hibernat-ing aquatic frogs, however, must be near oxygen-rich water and spend a good portion of the winter just lying on top of the mud or only partially buried. They may even slowly swim around from time to time.

Terrestrial frogs normally hibernate on land. American toads and other frogs that are good diggers burrow deep into the soil, safely below the frost line. Some frogs, such as the wood frog and the spring peeper, are not adept at dig-ging and instead seek out deep cracks and crevices in logs or rocks, or they just dig down as far as they can in the leaf litter. These hibernacula are not as well protected from frigid weather and may freeze, along with their inhabi-tants.

And yet the frogs do not die. Why? Antifreeze! True enough, ice crystals form in such places as the body cav-ity and bladder and under the skin, but a high concentra-tion of glucose in the frog's vital organs prevents freezing. A partially frozen frog will stop breathing, and its heart will stop beating. It will appear quite dead. But when the hibernaculum warms up above freezing, the frog's frozen portions will thaw, and its heart and lungs resume activity—there really is such a thing as the living dead!

Estivation is similar to hibernation. It is a dormant state an animal assumes in response to adverse environmental conditions, in this case, the prolonged dry season of certain tropical regions. Several species of frog are known to estivate. Two of the better-known species are the ornate horned frog from South America and the African bullfrog.

When the dry season starts, these frogs burrow into the soil and become dormant. During the extended dry season, which can last several months, they perform a neat trick: They shed several intact layers of skin, forming a virtually waterproof cocoon that envelops the entire body, leaving only the nostrils exposed, which allows them to breathe. They remain in their cocoons for the duration of the dry season. When the rains return, the frogs free themselves of their shrouds and make their way up through the moist soil to the surface.

?

Do unbred animals lack the individual distinctiveness of humans?

ANSWERED BY:
Pamela J. Pietz, Northern Prairie Science Center, Jamestown, North Dakota.

Humans can tell individual animals apart in many species, but it takes some familiarity with the species of interest. For example, researchers can recognize individual humpback whales because each whale has a unique black-and-white pattern on the underside of its tail flukes.

It is likely that we could recognize individuals of many species if we spent enough time observing them carefully. Of course, individuals of some species do look more alike than individuals of some other species. For example, some invertebrates have reproductive systems that lead to many individuals being very closely related to one another genetically (much like identical twins in humans). The less genetic variability there is among individuals of a species, the more closely they will resemble one another.

We should also remember that visual markers are not

the only possible way to tell individuals apart. Humans tend to rely on visual cues more than other types of cues because our vision is more highly developed than our other senses.

Individual animals of a given species probably can tell one another apart as easily as we can tell humans apart, but they may use sound, smell, and other senses instead of, or in addition to, vision. Birds are strongly visually oriented (that's why they are so colorful), so they may use visual cues to recognize individuals; they also have excellent hearing, however, so they may respond to differences in individual voices as well, much as humans do. Reptiles use chemical signs (akin to smell) to gather information about their environment, so they probably likewise rely on chemical signs to tell individuals apart. Some animals see beyond the visible light spectrum (bees and some birds see ultraviolet wavelengths), and some animals hear sounds that are too low (e.g., elephants) or too high (e.g., dogs) for humans to hear. Thus, some animals may use cues to tell one another apart that are not available to us.

Talkin' About Evolution
evolution

?

Is there any evolutionary advantage to gigantism?

ANSWERED BY:

Gregory S. Paul, Freelance Scientist and Artist, and Editor of The Scientific American Book of Dinosaurs.

Gigantism applies to animals that exceed 1 ton. Today's land giants include elephants (which weigh between 5 and 10 tons), rhinos, hippos, and giraffes. Yet these creatures represent only a small percentage of the terrestrial giants that have existed in the Mesozoic era and Cenozoic era (which we are still in). Indeed, gigantism has been a common feature of land animals since the beginning of the Jurassic period, over 200 million years ago. The first true giants of the land were small-headed, long-necked, sauropod dinosaurs that appeared at the beginning of that time. Soon some of the predatory dinosaurs exceeded 1 ton. Toward the end of the Jurassic, many sauropods reached 10 to 20 tons, some weighed as much as 50 tons, and a few may have exceeded 100 tons and 150 feet in length, rivaling the largest modern whales.

The bigger an animal is, the safer it is from predators and the better it is able to kill prey. Antelope are easy prey for lions, hyenas, and hunting dogs, but adult elephants and rhinos are nearly immune—and their young benefit from the protection of their huge parents. For herbivores, being gigantic means being taller and therefore able to access higher foliage. Giraffes and elephants can reach over 18 feet high, and elephants can use their great bulk to push over even taller trees.

Other reasons for being gargantuan are less obvious, although important. The cost of locomotion decreases with increasing size; thus it is much cheaper for a five-ton elephant to walk a mile than it is for a five-ton herd of gazelle to move the same distance. Metabolic rate also decreases with increasing size. A shrew must therefore frantically eat more than its own weight each day. The elephant, on the other hand, needs to take in only five percent of its own weight. And whereas big herbivores have long digestive systems that allow them to process and digest tougher plants, small herbivores can survive only on higher-quality foods. Also, as size increases, great bulk acts as a form of mass insulation. Large animals are thereby less affected by temperature extremes.

But there are disadvantages to being big. Because big animals eat more, there cannot be as many of them. Before human hunting, the population of elephants and rhinos in Africa was in the low millions. Rodents, in contrast, number in the countless billions. Nor can giants do a lot of things that smaller creatures can do, such as burrow into the ground, climb trees, or fly.

On land only dinosaurs and mammals have become gigantic; reptiles have never done so (the biggest tortoises and lizards have only weighed a ton). One reason may be the rate of growth. Land reptiles cannot grow rapidly. It takes many years for an alligator to reach 100 pounds, whereas an ostrich does so in less than a year. This is probably related to metabolic rate. Only those animals with high metabolic rates can grow rapidly, so only they can grow fast enough to become gigantic within a reasonable amount of time. Elephants reach as much as 10 tons in just three decades. It would take well over a century for a reptile to do the same.

It is widely believed that it is more logical for giants to dwell in oceans, where they are buoyed by water, than on land, where gravity is a force that must be constantly resisted. Yet sauropods were the most successful giants in the earth's history, having survived for 150 million years—the 100-tonners themselves lasted for 80 million years. Gigantic whales, in contrast, have swum the seas for only 40 million years, and only the blue whale has reached 100 tons.

?

What is the point in preserving endangered species that have no practical use to humans?

ANSWERED BY:
Kerry Bruce Clark, Professor of Biological Sciences, Florida Institute of Technology, Melbourne.

E very organism, whether or not it has direct practical use to humans, has a functional role (or niche) in its habitat or ecosystem. Though many species appear to have trivial niches, we should remember that the relative effects of various organisms in biological systems are seldom static; and minor species can sometimes become very important as systems fluctuate. Each species also represents a unique genetic library. Our genetic technology is only beginning to tap the vast potential benefits of these libraries, and seemingly minor species are typically the most specialized organisms; we can expect that ecological specialists will often turn out to have the most unusual genes and hence represent potential resources that we should preserve for our future needs.

Minor species often have functions that we may not understand but that may be ecologically or evolutionarily important, often involving complex interactions of many other species, some of which may in turn be ecologically or commercially important. The dodo and the Carolina parakeet were important dispersers of seeds, and their loss has permanently affected forest structure in their habitats; rare insects are often highly specific pollinators whose loss affects the reproduction and survival of other plants. On evolutionary time scales, we know far less about the effects of extinction of rare species, but we do know that evolution can amplify the effect of a species over time through its interactions on the survival of other species. In most cases, we simply do not know enough about the biology of a rare species to predict the effects of its extinction. But once the species is lost, we can never provide a perfect substitute.

When we lose one rare species, it actually symbolizes many changes of far broader impact, ranging from the loss of habitats (affecting large numbers of species) to large-scale alterations to the functions of those habitats. As the human population climbs, these cumulative changes will ultimately affect our economies and our well-being, because natural ecosystems perform—free of charge— many functions that we take for granted, such as purification of our wastes, production of harvestable resources, regulation of our climate, and restoration of the oxygen we breathe.

?

What do we know about the evolution of sleep?

ANSWERED BY:
Irene Tobler, Researcher, Institute of Pharmacology, University of Zurich, Switzerland.

M ost animals, and probably most living organisms, exhibit a circadian rest-activity rhythm. It is possible that sleep may have evolved from rest to allow more flexibility within this rather rigid rhythm of rest and activity. Researchers think that sleep arose to allow organisms to conserve and restore their energy. All mammals and birds sleep, as defined by the typical changes in brain waves associated with sleep. Reptiles also display changes in brain-wave states that correlate with changes in behavior, but scientists have not unambiguously demonstrated the existence in reptiles of the two sleep states (rapid-eye-movement sleep and non-rapid-eye-movement sleep) that are found in mammals and birds. For amphibia and fish, it is possible to make a behavioral definition of sleep-like states, although not many species have been investigated.

For invertebrates, it is possible to identify elements—such as body position, arousal threshold, muscle activity, and heart rate—that allow us to differentiate rest from sleep. Particularly in insects, but also in the scorpion and some crustaceans, researchers have identified a rest state that is similar to sleep.

In addition to a circadian component, sleep in vertebrates is characterized by a homeostatic component: Sleep is regulated in its intensity as a function of the duration of previous wakefulness. Experiments have elicited a homeostatic response to some hours of "rest deprivation" in scorpions and several cockroach species—that is, they were more restful during recovery. This finding indicates that regulatory aspects of sleep may be present in invertebrates as well. Further back in the evolutionary tree, the existence of a sleep state has been described only anecdotally (for example, in cephalopods—squids and octopi—and aplysia—"sea hair" mollusks). The presence of a true sleep state among these creatures remains to be investigated scientifically.

When Dinosaurs Ruled the Earth
dinosaurs

?

What are the odds of a dead dinosaur becoming fossilized?

ANSWERED BY:
Gregory M. Erickson, Paleontologist, Florida State University, Tallahassee.

I t is often stated in the paleontological literature that the chance an animal will become fossilized is "one in a million." This number is meant to be taken figuratively, the point being that the odds of surviving the rigors of deep time are extremely remote. Nevertheless, all field paleontologists know that the earth is biased when it comes to giving up its dead—the odds of an animal being preserved and consequently exhumed are much greater in some settings than others.

Studies have shown that organisms that die on land in lush jungle locales are rarely fossilized. In these settings, there is little chance of being buried, scavenging vertebrates and insects are prevalent, bacteria that break down flesh and bones are abundant, and the soils are extremely acidic and tend to dissolve bones. As a result, remains of dinosaurs from former surroundings are practically nonexistent. Conversely, dinosaurs are commonly found in areas that were once fluvial (river) settings and in regions of extreme aridity. In the former case, it is clear that dinosaur remains were rapidly buried before substantial scavenging could take place. Remains of dinosaurs that were washed into the fluvial systems are found buried in actual river channels, whereas others are discovered on the former floodplains at the location where they fell and were covered by sediments from floodwaters that breached riverbanks. Because river currents tend to scatter and break up bones, remains from river channels are often biased toward certain bones depending on the strength of the current.

Once bones were entombed in fluvial sediments, not only were they protected from scavengers and many types of bioorganisms, but they could also begin a process known as permineralization. Water percolating through the sands or muds was often rich in silica (natural glass) and other minerals, which could fill in the pores of the bones and make them physically resistant to crushing by the overlying sediment.

Dinosaurs dying in arid regions also stood a reasonable chance of becoming fossilized, because aridity tends to desiccate a carcass, making it less attractive to scavengers. And unlike jungle or forest settings, deserts have considerably fewer organisms suited for the breakdown of animal tissues. Windblown sands, as well as drifting and collapsing sand dunes, were agents of burial for such animals. Subsequent rainfall during the wet seasons carried minerals into the buried bones.

If dinosaur remains entombed in the ways described above did not later become modified by upheavals of the earth, then there is a good chance they are still around today.

?

What kind of evidence could be found in the fossil record (or anywhere else) that would prove whether some dinosaurs were warm-blooded?

ANSWERED BY:
Thomas E. Williamson, New Mexico Museum of Natural History and Science, Albuquerque.

As yet, there is probably no evidence that would definitively prove whether some dinosaurs were warm-blooded. Scientists have explored numerous lines of evidence to try to answer this question.

There is a clear difference in bone structure between

modern cold-blooded and warm-blooded animals. Warm-blooded animals tend to have highly vascularized bone tissue. Cold-blooded animals, on the other hand, have relatively dense bone, sometimes even showing annual growth rings. Dinosaurs tend to have highly vascularized bone early in life and then develop dense bone with growth rings as they reach maturity. New evidence suggests that the different bone types are more related to growth rates than to warm- or cold-bloodedness.

In modern animal communities, there are far fewer warm-blooded predators relative to prey than cold-blooded predators. This is because warm-blooded animals tend to eat far more than cold-blooded animals and so a given amount of prey will support far fewer warm-blooded predators. Unfortunately, it is very difficult to determine the actual relative abundance of predator versus prey from the fossils because the fossil record does not always preserve an accurate representation of the original animal community.

Modern warm-blooded animals tend to have more erect postures than cold-blooded animals. Most dinosaurs have erect postures and therefore it has been suggested that this indicates that they had high activity levels and were warm-blooded. Along these lines, scientists have looked indirectly at the potential blood pressures of dinosaurs; warm-blooded animals tend to have relatively high blood pressures. Blood pressure can be estimated by looking at the vertical distance between the head and the heart. For some sauropod dinosaurs, this estimated blood pressure is very high indeed. The brachiosaurus would have had a blood pressure of about 500 millimeters of mercury. This figure is about five times higher than that of a human. On the other hand, other dinosaurs—such as ceratopsians—would have had a very low blood pressure, closer to that of living reptiles, based on this analysis.

Modern warm-blooded animals have relatively larger brains than living cold-blooded animals. It is thought that large brains are needed to coordinate active, highly energetic animals. Most dinosaurs have very small brains relative to their body size. In fact, their brains closely resemble those of modern reptiles. A few small, predatory dinosaurs have relative brain sizes that are comparable with those of some living birds, such as ostriches, however.

There is evidence from dinosaur trackways, mass accumulations of certain dinosaur fossils, and nesting sites that dinosaurs were social animals. Some have argued that such sophisticated behavior is suggestive of warm-bloodedness.

Birds are warm-blooded and probably evolved from a group of meat-eating dinosaurs. Therefore, it has been argued, their dinosaur ancestors were also warm-blooded. But recent study of the bone structure of some of the earliest birds has revealed that it resembles that of modern cold-blooded reptiles, suggesting that the first birds were cold-blooded and that warm-bloodedness developed later.

More recently, some researchers have looked for respiratory turbinates in the nasal passages of dinosaurs. Respiratory turbinates are fine, scroll-shaped bones found in the noses of most modern warm-blooded animals (mammals and birds). These structures are believed to function as moisture-recovery organs, recapturing water from warm and moist exhaled air. Without these structures, many warm-blooded animals would quickly dehydrate, especially in dry climates. So far, no respiratory turbinate structures have been found in dinosaurs. It has been argued, however, that many dinosaurs lived in warm and moist environments where water conservation may not have been important.

It is quite possible that dinosaurs had a metabolism that is different from that of living animals. Indeed, the large size attained by many dinosaurs may have led to what has been called "inertial homeothermy" or "gigantothermy." That is, the large bulk of the animals would have allowed them to maintain a fairly constant temperature even without a high metabolism. Certainly, dinosaurs were active and, in some cases, social animals that successfully competed with warm-blooded mammals during their 160-million-yearlong reign on the earth.

?

How close are we to being able to clone a dinosaur?

ANSWERED BY:

Jack Horner, Paleontologist, Montana State University, Bozeman, Montana, and Curator, Museum of the Rockies.

We are a long, long way from being able to reconstruct the DNA of extinct creatures, and in fact it may be impossible to resurrect the DNA of dinosaurs or other long-extinct forms. We have DNA for living creatures, including ourselves, and yet we cannot clone any living animal (from DNA alone). As for extinct animals, it is unclear whether DNA actually can survive more than a few thousand years. No one has been able to demonstrate incontrovertibly, as of yet, that they can retrieve DNA from an extinct species.

The cloning that has been achieved recently has been accomplished at the cellular level, not at the level of a strand of DNA. And to date we do not have any living cells of any extinct animals. Without cells, we cannot accomplish the same kind of cloning that has been done with sheep; and without being able to acquire ancient DNA or being able to clone from DNA, it is at present not possible even to forecast when such a thing might be possible. And even if we could get DNA from extinct animals, and if we knew how to clone the DNA, we would still have another hurdle in creating the exact embryonic conditions.

?

Did any dinosaurs have poisonous saliva, as in *Jurassic Park*?

ANSWERED BY:
Scott D. Sampson, Curator of Vertebrate Paleontology, Utah Museum of Natural History, and Assistant Professor of Geology and Geophysics, University of Utah, Salt Lake City.

There is no conclusive evidence that any dinosaur possessed poisonous saliva, and the poison-spitting dinosaur depicted in *Jurassic Park* was not based on any real evidence. However, the Mexican paleontologist Rubén A. Rodríguez de la Rosa of the Museum of the Desert in Saltillo has recently recovered a single odd tooth from an unknown carnivorous dinosaur that possesses a groove reminiscent of those seen in snakes for transmitting poison. I have examined this tooth firsthand and, at this point, remain unconvinced. More examples need to be recovered before this hypothesis can be evaluated.

That said, it is well known that the largest lizard predator today, the Komodo dragon, does carry a potentially lethal bacterial load in its saliva that is used to poison prey. Not surprisingly, several people have suggested that some carnivorous dinosaurs may have exhibited similar and equally unsavory behaviors, biting prey and letting the poison do the rest of the work. This might have proved an especially useful tactic for Jurassic predators such as the allosaurus, which most likely tackled prey animals many times its body size: the gargantuan, long-necked sauropods, such as the brachiosaurus and the apatosaurus. Nevertheless, it is important to note that all of the above remains firmly embedded in speculation.

If T. rex fell, how did it get up, given its tiny arms and low center of gravity?

ANSWERED BY:
Gregory M. Erickson, Paleontologist, Florida State University, Tallahassee.

Scientific inquiry has focused on the utility of the diminutive arms of tyrannosaurs for nearly a century. Several theories, including some regarding the arms' role in raising these animals from the ground, have long been kicked around. The American paleontologist Henry Fairfield Osborn, the first one to describe T. rex, initially expressed doubts that the relatively small humerus, or upper arm bone, associated with this enormous animal really belonged to it. Once convinced, however, he forwarded the first theory in 1906 of their utility—in grasping organs for copulation. In 1970 the British paleontologist Barney Newman posited that the arms actually served as braces to prevent the front of the body from skidding forward as the animal rose from a prone position using its hindlimbs. During such activity, the forelimbs would have been extended in an action reminiscent of a push-up. Other competing theories contend that the arms are vestigial (degenerate organs that have lost much use) or that they functioned as meat hooks while the creature's teeth were employed.

Are any of these theories correct? We may never know the answer. Nevertheless the recent finding of the first specimens of complete T. rex forelimbs in northern Montana has opened the door to biomechanical analyses and osteopathic observations from which new insights into the physical capacities of these structures have emerged. It is now clear that T. rex's hands could not reach its mouth. The elbow could not be extended much beyond a 90-

degree angle. The arms were very strong (perhaps capable of curling nearly 400 pounds) but had a very limited range of motion, both side-to-side and up-and-down. The wrists were considerably weaker and do not seem suited for supporting large mechanical loads. The small T. rex arms were often broken during life, which suggests that they were poorly suited for whatever the dinosaurs were trying to use them for and, more importantly, that these animals could go without using their arms for periods of up to a month.

Collectively, these findings seem to fly in the face of just one of the aforementioned theories: Newman's push-up theory. If this is the case, then how did T. rex get up? We can look to the birds (avian dinosaurs) for the answer, as they can stand up without the aid of arms. It is simply a matter of getting one's limbs below the center of gravity before extending them. I am not aware of any studies suggesting that tyrannosaurs could not do this. Furthermore, tyrannosaurs would have had the additional aid of their tails. From skeletal evidence and preserved trackways from the related albertosaur (in which the tails did not drag), it is clear that tyrannosaur tails acted as counterbalances—10,000-pound walking, teeter-totters. The tail would have helped to keep the center of balance back on the body as the hindlimbs were moved into position underneath.

Clearly tyrannosaurs got up at least once during their lives (at birth), and there is no reason to believe they could not throughout life—armed with pathetic arms or not.

3

Being
Human

It's All in the Genes
human evolution

?

Is the human race still evolving? Isn't culture
a more powerful force?

ANSWERED BY:

Meredith F. Small, Associate Professor, Department of
Anthropology, Cornell University, Ithaca, New York.

First of all, humans haven't really changed the rules of natural selection. We might think that because we have culture—and with it all kinds of medical interventions and technologies—that we are immune from natural selection, but nature proceeds as usual. Evolution is defined as a change in gene frequencies over time, which means that over generations, there will be changes in the gene pool, and humans experience those changes as much as any other organism. Some people live, and some people die, and some people pass on more genes than others. Therefore, there is a change in the human gene pool over time.

But we might suggest that, with all that cultural and technological intervention, there would be some kind of influence in the composition of the gene pool, and there is. Take smallpox, for an example. Years ago millions of people died from smallpox, and their genes were not passed on because many of them died before reproductive age. The human gene pool was then missing the genes of those people. But now, since smallpox has been wiped off the planet, people who would have normally died of the disease now live, probably have children, and thus contribute to the human gene pool. In another example, the birth rate always goes down the more developed and economically affluent countries become. Today the highest birth rates are in Latin America, Africa, and Asia. People in these places are now the major contributors to the human gene pool. In many generations, the human species will be composed more of genes from those groups than from developed countries.

And so culture, development, and medicine might change the tenor of the human gene pool, but they do not take away the force of evolution. Also, keep in mind that culture may not seem a natural force, but because it is part of our environment, it is just as natural as disease, weather, or food resources. We in developed nations may think we are immune from natural selection because we are so surrounded by material goods and high technology, but this immunity is an illusion. Technology protects us from nothing, and medicine surely hasn't cured all the diseases. We in developed nations are more comfortable, but we still die, and we still contribute differentially to future generations. And most important, we have to realize that the developed-nation view of the human species is a very narrow take on humanity. The majority of the human population does not live like this; more than half the people on the earth have never spoken on a telephone.

?

Can the human race be devolving?

ANSWERED BY:
Michael J. Dougherty, Assistant Director and Senior Staff Biologist, Biological Sciences Curriculum Study, Colorado Springs, Colorado.

From a biological perspective, there is no such thing as devolution. All changes in the gene frequencies of populations—and quite often in the traits those genes influence—are by definition evolutionary changes. The notion that humans might regress or "devolve" presumes that there is a preferred hierarchy of structure and function—say, that legs with feet are better than legs with hooves or that breathing with lungs is better than breathing with gills. But for the organisms possessing those structures, each is a useful adaptation.

Many people evaluate nonhuman organisms according to human anatomy and physiology and mistakenly conclude that humans are the ultimate product, even goal, of evolution. That attitude probably stems from the tendency of humans to think anthropocentrically. Unfortunately, anthropocentric thinking is at the root of many common misconceptions in biology.

Chief among these misconceptions is that species evolve or change because they need to change to adapt to shifting environmental demands; biologists refer to this fallacy as teleology. In fact, more than 99 percent of all species that ever lived are extinct, so clearly there is no requirement that species always adapt successfully. As the fossil record demonstrates, extinction is a perfectly natural—and indeed quite common—response to changing environmental conditions. When species do evolve, it is not out of need but rather because their populations contain organisms with variants of traits that offer a reproductive advantage in a changing environment.

Another misconception is that increasing complexity is the necessary outcome of evolution. In fact, decreasing complexity is common in the record of evolution. For example, the lower jaw in vertebrates shows decreasing complexity, as measured by the numbers of bones, from fish to reptiles to mammals. (Evolution adapted the extra jawbones into ear bones.) Likewise, ancestral horses had several toes on each foot; modern horses have a single toe with a hoof.

Evolution, not devolution, selected for those adaptations.

?

Why are we getting taller as a species?

ANSWERED BY:
Michael J. Dougherty, Assistant Director and Senior
Staff Biologist, Biological Sciences Curriculum Study,
Colorado Springs, Colorado.

Anyone who has ever visited a home built around the time of the Revolutionary War along the back alleys of Philadelphia or Boston has been struck, metaphorically if not literally, by the characteristically low ceilings and small door frames. Even houses built in the early 1800s can make a person of average height by today's standards wonder how the original occupants managed to stay conscious long enough to participate in an industrial revolution and a civil war.

For most people, contemporary buildings do not prompt similar claustrophobic concerns. The reason for this difference, as many people have correctly guessed, is that modern humans are taller than those from the eighteenth and nineteenth centuries. In fact, over the last 150 years the average height of people in industrialized nations has increased approximately 10 centimeters (about four inches).

Why this relatively sudden growth? Most geneticists believe that the improvement in childhood nutrition has been the most important factor in allowing humans to increase so dramatically in stature. The evidence for this argument is threefold.

First, the increase in height has not been continuous since the dawn of man; it began sometime around the middle of the nineteenth century. In fact, examinations of skeletons show no significant differences in height from the stone age through the early 1800s. Also, during World Wars I and II, when hunger was a frequent companion of the German civilian population, the heights of the children actually declined. They only recovered during the postwar years.

Second, the trend toward increasing height has largely leveled off, suggesting that there is an upper limit to height beyond which our genes are not equipped to take us, regardless of environmental improvements. Interestingly, the age of menarche, which is also influenced by nutrition, has shown a corresponding decrease over this same time period. Some scientists believe that the increase in teenage and out-of-wedlock pregnancies in the developed world may be an unanticipated consequence of improved nutrition.

Third, conditions of poor nutrition are well correlated to smaller stature. For example, the heights of all classes of people, from factory workers to the rich, increased as food quality, production, and distribution became more reliable, although class differences still remain. Even more dramatic, the heights of vagrant London boys declined from 1780 to 1800 and then rose three inches in just 30 years—an increase that paralleled improving conditions for the poor. Even today, height is used in some countries as an indicator of socioeconomic division; and differences can reveal discrimination within social, ethnic, economic, occupational, and geographic groups.

For those hoping that humans might someday shoot basketballs through 15-foot-high hoops, the fact that the increase in human height is leveling off no doubt will be disappointing. For those who understand, however, that our genes are merely a blueprint that specifies what is possible given an optimal environment, a limit on height is just one of many limitations in life and certainly not the most constraining.

Why do men have nipples?

ANSWERED BY:
**Andrew M. Simons, Professor of Biology,
Carleton University, Ottawa, Canada.**

Evolutionary biologists, whose job it is to explain variety in nature, are often expected to provide adaptive explanations for such "why" questions. Some traits may prove—through appropriate tests—to be best explained as adaptations; others have perfectly good evolutionary, but nonadaptive, explanations. This is because evolution is a process constrained by many factors including history, chance, and the mechanisms of heredity, which also explains why particular attributes of organisms are not as they would be had they been "designed" from scratch. Nipples in male mammals illustrate a constrained evolutionary result.

A human baby inherits one copy of every gene from his or her father and one copy of every gene from his or her mother. Inherited traits of a boy should thus be a combination of traits from both his parents. Thus, from a genetic perspective, the question should be turned around: How can males and females ever diverge if genes from both parents are inherited? We know that consistent differences between males and females (so-called sexual dimorphisms) are common—examples include bird plumage coloration and size dimorphism in insects. The only way such differences can evolve is if the same trait (color, for example) in males and females has become "uncoupled" at the genetic level. This happens if a trait is influenced by different genes in males and females, if it is under control of genes located on the sex chromosomes, or if gene expression has evolved to be dependent on context (whether genes find themselves within a male or a female genome). The idea of the shared genetic basis of two traits (in this case in males and females) is known as a genetic correlation, and it is a quantity routinely measured by evolutionary geneticists. The evolutionary default is for males and females to share characters through genetic correlations.

The uncoupling of male and female traits occurs if there is selection for it: if the trait is important to the reproductive success of both males and females but the best or "optimal" trait is different for a male and a female. We would not expect such an uncoupling if the attribute is important in both sexes and the "optimal" value is similar in both sexes; nor would we expect uncoupling to evolve if the attribute is important to one sex but unimportant in the other. The latter is the case for nipples. Their advantage in females, in terms of reproductive success, is clear. But because the genetic "default" is for males and females to share characters, the presence of nipples in males is probably best explained as a genetic correlation that persists through lack of selection against them, rather than selection for them. In a sense, male nipples are analogous to vestigial structures such as the remnants of useless pelvic bones in whales: If they did much harm, they would have disappeared.

In a famous paper, Stephen Jay Gould and Richard C. Lewontin emphasize that we should not immediately assume that every trait has an adaptive explanation. Just as the spandrels of St. Mark's domed cathedral in Venice are simply an architectural consequence of the meeting of a vaulted ceiling with its supporting pillars, the presence of nipples in male mammals is a genetic architectural by-product of nipples in females. So, why do men have nipples? Because females do.

Oh, Behave!
human behavior

?

How did the smile become a friendly gesture in humans?

ANSWERED BY:
**Frank McAndrew, Professor of Psychology,
Knox College, Galesburg, Illinois.**

Baring one's teeth is not always a threat. In primates, showing the teeth, especially teeth held together, is almost always a sign of submission. The human smile probably has evolved from that.

In the primate threat, the lips are curled back and the teeth are apart—you are ready to bite. But if the teeth are pressed together and the lips are relaxed, then clearly you are not prepared to do any damage. These displays are combined with other facial gestures to express a whole range of feelings. In a lot of human smiling, it is something you do in public, but it does not reflect true "friendly" feelings—think of politicians smiling for photographers.

What is especially interesting is that this is not learned but preprogrammed behavior. Children who are born blind never see anybody smile, but they show the same kinds of smiles under the same situations as sighted people.

?

Why are more people right-handed? Do other primates show a similar tendency to favor one hand over the other?

ANSWERED BY:

M. K. Holder, Affiliated Scientist, Center for the Integrative Study of Animal Behavior, Indiana University, Bloomington.

In the 160 years in which "handedness" has been studied, we have learned quite a lot, but we still cannot precisely describe what causes humans preferentially to use one hand over the other or why human populations are biased toward right-hand use rather than left-hand use.

Scientists disagree over what percentage of human populations are right-handed or left-handed because there is no standard definition for measuring handedness; our criteria vary and are based on various theoretical explanations because we are still trying to understand the mechanisms involved. But we can describe in general terms what we do know.

Most humans (say 70 percent to 95 percent) are right-handed; a minority (say 5 percent to 30 percent) are left-handed; and an indeterminate number of people are probably best described as ambidextrous. This appears to be universally true for all human populations anywhere in the world. There is evidence for genetic influence for handedness; however, geneticists cannot agree on the exact process. There is evidence that handedness can be influenced (and changed) by social and cultural mechanisms. For instance, teachers have been known to force children to switch from using their left hand to using their right hand for writing. Also, some more restrictive societies show less left-handedness in their populations than other more permissive societies.

Some researchers argue there is evidence for cases of "pathological" left-handedness related to brain trauma during birth. And many researchers trace the cause of handedness back to prenatal developmental processes, to the time when the fetal brain is first developing distinct cerebral hemispheres. In the 1860s the French surgeon Paul Broca noted a relationship between right-handedness and left-hemispheric brain specialization for language abilities. But the hand–brain association is neither a simple, nor reliable, correlation. Studies conducted in the 1970s showed that most left-handers have the same left-hemispheric brain specialization for language typical of all humans—only a portion of left-handers have different patterns of language specialization.

So the bottom line is, we have a good general idea of the causes of right-handedness in human populations, but we have yet to work out the precise details, including why the direction is right instead of left.

The second question is currently a controversial one. It is important to note the difference between an individual animal being left- or right-handed and most of the animals in an entire population being either left- or right-handed. It is not unusual for individual animals to show a preferential use of one hand over the other. But there is no consensus among researchers that any nonhuman species shows the same species-level handedness found in humans.

There are a few researchers who argue for this, but most of them work with animals in laboratory or captive settings, performing manual tasks that are very different from how animals use their hands in the wild.

In addition to studying handedness in humans, I have also studied hand usage in mountain gorillas (in Rwanda) as well as chimpanzees, red colobus monkeys, redtail monkeys, and grey-cheeked mangabeys (in Uganda). My own research shows that individual monkeys and apes often develop individual preferences (both left and right) for manual tasks, but I have found no evidence for population-level handedness, as seen in humans.

?

How long can humans stay awake?

ANSWERED BY:
J. Christian Gillin, Professor of Psychiatry, University of California, San Diego.

The easy answer to this question is 264 hours (about 11 days). In 1965, Randy Gardner, a 17-year-old high school student, set this apparent world record for a science fair. Several other healthy research subjects have remained awake *for 8 to 10 days* in carefully monitored experiments. None of these individuals experienced serious medical, neurological, physiological, or psychiatric problems. On the other hand, all of them showed progressive and significant deficits in concentration, motivation, perception, and other higher mental processes as the duration of sleep deprivation increased. All of the experimenters recovered to relative normality within one or two nights of recovery sleep. Other anecdotal reports describe soldiers staying awake for four days in battle or unmedicated patients with mania going without sleep for three to four days.

The more difficult answer to this question revolves around the definition of "awake." Prolonged sleep deprivation in healthy subjects induces altered states of consciousness (often described as "microsleep"), numerous brief episodes of overwhelming sleep, and loss of cognitive and motor functions. We all know about the dangerous, drowsy driver, and we have heard about sleep-deprived British pilots who crashed their planes (having fallen asleep) while flying home from the war zone during World War II. Randy Gardner was "awake" but basically cognitively dysfunctional at the end of his ordeal.

In certain rare human medical disorders, the question of how long people can remain awake raises other surprising answers and more questions. Morvan's fibrillary chorea or Morvan's syndrome is characterized by muscle twitching, pain, excessive sweating, weight loss, periodic hallucinations, and severe loss of sleep (agrypnia). Michel Jouvet and his colleagues in Lyon, France, studied a 27-year-old man with this disorder and found he had virtually no sleep over a period of several months. During that time he did not feel sleepy or tired and did not show any disorders of mood, memory, or anxiety. Nevertheless, nearly every night between 9:00 and 11:00 P.M., he experienced a

20- to 60-minute period of auditory, visual, olfactory, and somesthetic (sense of touch) hallucinations.

To return to the original question, "How long can humans stay awake?" the ultimate answer remains unclear. I am unaware of any reports that sleep deprivation per se has killed any human (excluding accidents and so forth). Indeed, the U.S. Department of Defense has offered research funding for the goal of sustaining a fully awake, fully functional "24/7" soldier, sailor, or airman. Future warriors will face intense, around-the-clock fighting for weeks at a time. Will bioengineering eventually produce genetically cloned soldiers and citizens who need no sleep but remain effective and happy? I hope not. A good night's sleep is one of life's blessings.

Do humans have some kind of homing instinct like certain birds do?

ANSWERED BY:
**C. Randy Gallistel, Professor of Psychology,
University of California, Los Angeles.**

Researchers have conducted a number of experiments to determine whether humans have a magnetic compass sense, but these have been inconclusive. Expert opinion is fairly unanimous that there are no convincing indications of such a sense in humans. On the other hand, there is good evidence that many insects, birds, and reptiles have such a sense, although it remains unclear under which conditions they make use of it.

Having a magnetic compass sense is not equivalent to having a homing instinct, because knowing which way is north does you no good if you do not know whether you are north, south, east, or west of home. Many animals, including humans, keep track of where they are (and hence the direction to home) by a method known as dead reckoning: As they move about, they keep track of each individual movement, adding these up to derive their net change in position.

Dead reckoning is no help, however, when people or animals are displaced under conditions in which it is impossible for them to determine the speed and direction in which they are moving. Nevertheless, many animals—most notably homing pigeons—are able to figure out where they are even after this kind of displacement. How they do so remains a mystery, despite much experimental work on this phenomenon.

?

Why do we yawn when we are tired? And why does it seem to be contagious?

ANSWERED BY:

Mark A. W. Andrews, Associate Professor of Physiology and Director of the Independent Study Program, Lake Erie College of Osteopathic Medicine, Erie, Pennsylvania.

Yawning appears to be not only a sign of tiredness but also a much more general sign of changing conditions within the body. Studies have shown that we yawn when we are fatigued, as well as when we are awakening and during other times when our state of alertness is changing.

Yawning is characterized by a single deep inhalation (with the mouth open) and stretching of the muscles of the jaw and trunk. It occurs in many animals and involves interactions between the unconscious brain and the body.

For years it was thought that yawns served to bring in more air when low oxygen levels were sensed in the lungs by nearby tissue. We now know, however, that the lungs do not necessarily detect an oxygen deficit. Moreover, fetuses yawn in utero, even though their lungs are not yet ventilated. In addition, different regions of the brain control yawning and breathing. Low oxygen levels in the paraventricular nucleus (PVN) of the hypothalamus of the brain can induce yawning. Another hypothesis is that we yawn because we are tired or bored. But this, too, is probably not the case—the PVN also plays a role in penile erection, an event not typically associated with boredom.

It does appear that the PVN of the hypothalamus is, among other things, the "yawning center" of the brain. It contains a number of chemical messengers that can induce yawns, including dopamine, glycine, oxytocin, and adrenocorticotropic hormone (ACTH). ACTH, for one, surges at night and prior to awakening and elicits yawning and stretching in humans. Yawning also seems to require production of nitric oxide by specific neurons in the PVN. Once stimulated, the cells of the PVN activate cells of the brain stem and/or hippocampus, causing yawning. Yawning likewise appears to have a feedback component: If you stifle or prevent a yawn, the process is somewhat unsatisfying.

It is correct to say that yawns are contagious: Seeing, hearing, or thinking about yawning can trigger the event, but there is little understanding of why. Many theories have been presented over the years. Some evidence suggests that yawning is a means of communicating changing environmental or internal body conditions to others, possibly as a way to synchronize behavior. If this is the case, yawning in humans is most likely a vestigial mechanism that has lost its significance.

You Haven't Aged a Bit
growing older

?

Why does hair turn gray?

ANSWERED BY:
Laurence Meyer, Dermatologist, University of Utah, Salt Lake City.

The pigment in hair, as well as in the skin, is called melanin. There are two types of melanin: eumelanin, which is dark brown or black, and pheomelanin, which is reddish yellow. Both are made by a type of cell called a melanocyte that resides in the hair bulb and along the bottom of the outer layer of skin, or epidermis. The melanocytes pass this pigment to adjoining epidermal cells called keratinocytes, which produce the protein keratin—hair's chief component. When the keratinocytes undergo their scheduled death, they retain the melanin.

Thus, the pigment that is visible in the hair and in the skin lies in these dead keratinocyte bodies.

The control of this pigment production is complex, and somewhat different for skin and hair, but there are clear genetic factors. One is the recently identified MC1R gene. Alleles of this gene are associated with red hair in humans, cows, and many other species. (Pigment can also tie in with the hair cycle—that is, the process of a new hair growing and stopping at a preset length.)

Gray hair, then, is simply hair with less melanin, and white hair has no melanin at all. Genes control this lack of deposition of melanin, too. In some families, many members' hair turns white in their 20s. Generally speaking, among Caucasians, 50 percent are 50 percent gray by age 50. There is, however, wide variation. This number differs for other ethnic groups, again demonstrating the effect of genetic control.

Exactly how hair loses its pigment remains unclear. In the early stages of graying, the melanocytes are still present but inactive. Later on they seem to decrease in number. In general, this type of graying is not associated with any disease, although it can be associated with some autoimmune processes. But graying in a young adult is not itself a sign of any health problem.

?

Do people lose their senses of smell and taste
as they age?

ANSWERED BY:

Charles J. Wysocki, Neuroscientist, Monell Chemical
Senses Center, Philadelphia.

It's true that as people age they often complain about a
decrease in—or even the loss of—their ability to taste a
superb meal or appreciate a fine beverage. When people
eat, however, they tend to confuse or combine information from the tongue and mouth (the sense of taste, which
uses three nerves to send information to the brain) with
what is happening in the nose (the sense of smell, which
utilizes a different nerve input).

It's easy to demonstrate this confusion. Grab a hand-
ful of jellybeans of different flavors with one hand and
close your eyes. With your other hand, pinch your nose
closed. Now pop one of the jellybeans into your mouth
and chew, without letting go of your nose. Can you tell
what flavor went into your mouth? Probably not, but you
most likely experienced the sweetness of the jellybean.
Now let go of your nose. Voilà—the flavor makes its
appearance.

This phenomenon occurs because smell provides most of the information about the flavor. Chemicals from the jellybean, called odorants, are inhaled through the mouth and exhaled through the nose, where they interact with special receptor cells that transmit information about smell. (It's the reverse process that one experiences downwind from a pig farm or chocolate factory.) These odorants then interact with the receptor cells and initiate a series of events that are interpreted by the brain as a smell.

Estimates for the number of odorant molecules vary, but there are probably tens of thousands of them. Taste, in contrast, is limited to sweet, sour, bitter, salty, and umami (the taste of monosodium glutamate, or MSG).

The sense of smell diminishes with advancing age—much more so than the sensitivity to taste. This decrease may result from an accumulated loss of sensory cells in the nose. The loss may be perhaps as much as two-thirds of the original population of 10 million. Although the elderly are in general less sensitive than young people to the overall perception of the food they eat, there are exceptions: Some 90-year-olds may be more sensitive to smells than some 20-year-olds.

Anatomy 101
the human body

?

What is the function of the human appendix?

ANSWERED BY:
Loren G. Martin, Professor of Physiology, Oklahoma State University, Stillwater.

For years, the appendix was credited with very little physiological function. We now know, however, that the appendix serves an important role in the fetus and in young adults. Endocrine cells appear in the appendix of the human fetus at around the eleventh week of development. These endocrine cells of the fetal appendix have been shown to produce various biogenic amines and peptide hormones—compounds that assist with various biological control, or homeostatic, mechanisms. There had been little prior evidence of this or any other role of the appendix in animal research, because the appendix does not exist in domestic mammals.

Among adult humans, the appendix is now thought to be involved primarily in immune functions. Lymphoid tissue begins to accumulate in the appendix shortly after birth and reaches a peak between the second and third decades of life, decreasing rapidly thereafter and practically disappearing after the age of 60. During the early years of development, however, the appendix has been shown to function as a lymphoid organ, assisting with the maturation of B lymphocytes (one variety of white blood cell) and in the production of the class of antibodies known as immunoglobulin A (IgA) antibodies. Researchers have also shown that the appendix is involved in the production of molecules that help to direct the movement of lymphocytes to various other locations in the body.

In this context, the function of the appendix appears to be to expose white blood cells to the wide variety of antigens, or foreign substances, present in the gastrointestinal tract. Thus, the appendix probably helps to suppress potentially destructive humoral (blood- and lymph-borne) antibody responses while promoting local immunity. The appendix—like the tiny structures called Peyer's patches in other areas of the gastrointestinal tract—takes up antigens from the contents of the intestines and reacts to these contents. This local immune system plays a vital role in the physiological immune response and in the control of food, drug, microbial, or viral antigens. The connection between these local immune reactions and inflammatory bowel diseases, as well as autoimmune reactions in which the individual's own tissues are attacked by the immune system, is currently under investigation.

In the past, the appendix was often routinely removed and discarded during other abdominal surgeries to prevent any possibility of a later attack of appendicitis; the appendix is now spared in case it is needed later for reconstructive surgery if the urinary bladder is removed. In such surgery, a section of the intestine is formed into a replacement bladder, and the appendix is used to re-create a "sphincter muscle" so that the patient remains continent (able to retain urine). In addition, the appendix has been successfully fashioned into a makeshift replacement for a diseased urethra, allowing urine to flow from the kidneys to the bladder. As a result, the appendix, once regarded as a nonfunctional tissue, is now regarded as an important "back-up" that can be used in a variety of reconstructive surgical techniques. It is no longer routinely removed and discarded if it is healthy.

What makes the sound when we crack our knuckles?

ANSWERED BY:
Raymond Brodeur, Ergonomics Research Laboratory, Michigan State University, East Lansing.

To understand what happens when you "crack" your knuckles, or any other joint, first you need a little background about the nature of the joints of the body. The type of joints that you can most easily "pop" or "crack" are the diarthrodial joints. These are your most typical joints. They consist of two bones that contact each other at their cartilage surfaces; the cartilage surfaces are surrounded by a joint capsule. Inside the joint capsule is a lubricant, known as synovial fluid, which also serves as a source of nutrients for the cells that maintain the joint cartilage. In addition, the synovial fluid contains dissolved gases, including oxygen, nitrogen, and carbon dioxide.

The easiest joints to pop are the ones in your fingers. As the joint capsule stretches, its expansion is limited by a number of factors. When small forces are applied to the joint, one factor that limits the motion is the joint's volume, which is set by the amount of synovial fluid contained in the joint. The synovial fluid cannot expand unless the pressure inside the capsule drops to a point at which the dissolved gases can escape the solution; when the gases come out of the solution, they increase the volume and hence the mobility of the joint.

The cracking or popping sound is thought to be caused by the gases rapidly coming out of the solution, allowing the capsule to stretch a little farther. The stretching of the joint is soon thereafter limited by the length of the capsule. If you take an x-ray of the joint after cracking, you can see a gas bubble inside the joint. This gas increases the joint volume by 15 to 20 percent; it consists mostly (about 80 percent) of carbon dioxide. The joint cannot be cracked again until the gases have dissolved back into the synovial fluid, which explains why you cannot crack the same knuckle repeatedly.

But how can releasing such a small quantity of gas cause so much noise? There is no good answer for this question. Researchers have estimated the energy levels of the sound by using accelerometers to measure the vibrations caused during joint popping. The amounts of energy involved are very small, on the order of 0.1 millijoule per cubic millimeter. Studies have also shown that there are two sound peaks during knuckle cracking, but the causes of these peaks are unknown. It is likely that the first sound is related to the gas dissolving out of the solution, whereas the second sound is caused by the capsule reaching its length limit.

A common, related question is, Does popping a joint cause any damage? There are actually few scientific data available on this topic. One study found no correlation between knuckle cracking and osteoarthritis in the finger joints. Another study, however, showed that repetitive knuckle cracking may affect the soft tissue surrounding the joint. Also, the habit tends to cause an increase in hand swelling and a decrease in the grip strength of the hand.

Another source of popping and cracking sounds is the tendons and ligaments near the joint. Tendons must cross at least one joint in order to cause motion. But when a joint moves, the tendon's position with respect to the joint is forced to change. It is not uncommon for a tendon to shift to a slightly different position, followed by a sudden snap as the tendon returns to its original location with respect to the joint. These noises are often heard in the knee and ankle joints when standing up from a seated position or when walking up or down stairs.

?

Why does your stomach growl when you are hungry?

ANSWERED BY:
Mark A. W. Andrews, Associate Professor of Physiology and Associate Director of the Independent Study Program at Lake Erie College of Osteopathic Medicine, Erie, Pennsylvania.

The physiological origin of this "growling" involves muscular activity in the stomach and small intestines. Although such growling is commonly associated with hunger—when the stomach and intestines are empty of contents that would otherwise muffle the noise—such sounds can occur at any time.

In general, the gastrointestinal tract is a hollow tube that runs from the mouth to the anus with walls primarily composed of layers of smooth muscle. This muscle is nearly always active to some extent. When these walls squeeze to mix and propel food, gas, and fluids, rumbling noises may be heard. Such squeezing, called peristalsis, involves a ring of contraction moving toward the anus, a few inches at a time.

A rhythmic fluctuation of electrical potential in the smooth muscle cells, known as the basic electrical rhythm (BER), generates the waves of peristalsis. BER is the result of the inherent activity of the enteric nervous system found in the walls of the gut. The autonomic nervous system and hormonal factors also modulate this rhythm.

After the stomach and small intestines have been empty for about two hours, there is a reflex generation of waves of electrical activity (migrating myoelectric complexes, or MMCs) in the enteric nervous system. These trigger hunger contractions, which can be heard as they clear out any stomach contents and keep them from accumulating at any one site.

?

How can you live without one of your kidneys?

ANSWERED BY:

Mark A. W. Andrews, Associate Professor of Physiology and Associate Director of the Independent Study Program, Lake Erie College of Osteopathic Medicine, Erie, Pennsylvania.

This is an excellent question, especially because kidney disease and kidney transplants are so common (approximately 10,000 to 15,000 Americans receive kidney transplants each year). Most humans are born with two kidneys as the functional components of what is called the renal system, which also includes two ureters, a bladder, and a urethra. The kidneys have many functions, including regulating blood pressure, producing red blood cells, activating vitamin D, and producing some glucose. Most evidently, however, the kidneys filter body fluids via the bloodstream to regulate and optimize their amount, composition, pH, and osmotic pressure. Excess water, electrolytes, nitrogen, and other wastes get excreted as urine. These functions maintain and optimize the "milieu interieur" (internal environment) of the body—the fluids in which our cells live.

Life is incompatible with a lack of kidney function. But unlike the case with most other organs, we are born with an overabundant—or overengineered—kidney capacity. Indeed, a single kidney with only 75 percent of its functional capacity can sustain life very well.

This overengineering supplies us with 1.2 million of the basic functional filtering element, the microscopic nephron, in each kidney. Nephrons are tiny tubes that filter the blood plasma, adjust, and then return optimized fluid to the body. Under most conditions, though totaling only a few pounds, the kidneys receive about 20 percent of all the blood pumped from the heart. Each day, about 120 liters of fluid and particles enter into the nephrons to be filtered.

If only one kidney is present, that kidney can adjust to filter as much as two kidneys would normally. In such a situation, the nephrons compensate individually by increasing in size—a process known as hypertrophy—to handle the extra load. This happens with no adverse effects, even over years. In fact, if one functional kidney is missing from birth, the other kidney can grow to reach a size similar to the combined weight of two kidneys (about one pound).

The kidneys filter this large amount of fluid on a daily basis because nephrons are fairly indiscriminate filters, removing all contents from the blood except for larger proteins and cells. The nephrons, however, are extremely accomplished in processing the filtrate and substances critical to survival—such as water, glucose, amino acids, and electrolytes, which are actively reabsorbed into the blood.

The water and waste (including urea and creatinine, acids, bases, toxins, and drug metabolites) that remain in the nephrons become urine.

In addition to being able to support life with only one kidney, the renal system has other safeguards. Although nephrons stop functioning at a rate of one percent per year after 40 years of age, the remaining nephrons tend to enlarge and fully compensate for this demise. Evidence strongly suggests that living kidney donors are highly unlikely to develop significant long-term detrimental effects to their health, as illustrated by donors whose renal function has been assessed for up to 30 years following donation. The main problems with donors are rare instances of complications having to do with the surgery, not the lack of the kidney.

?

Why do fingers wrinkle in the bath?

ANSWERED BY:
Laurence Meyer, Dermatologist, University of Utah, Salt Lake City.

The epidermis, or outer layer of the skin, is made up of cells called keratinocytes, which form a very strong intracellular skeleton made up of a protein called keratin. These cells divide rapidly at the bottom of the epidermis, pushing the higher cells upward. After migrating about halfway from the bottom of this layer to the top, the cells undergo a programmed death. The nucleus spirals inward, leaving alternating layers of the cell membrane, made of lipids, and the insides, made largely of water-loving keratin. The outer layer of the epidermis, called the stratum corneum, is thus composed of these alternating bands.

When hands are soaked in water, the keratin absorbs it and swells. The inside of the fingers, however, does not swell. As a result, there is relatively too much stratum corneum and it wrinkles, just like a gathered skirt. This bunching up occurs on fingers and toes because the epidermis is much thicker on the hands and feet than elsewhere on the body. The hair and nails, which contain different types of keratin, also absorb some water. This is why the nails get softer after bathing or doing the dishes.

Soaking in the tub does hydrate the skin but only briefly. All the added water quickly evaporates, leaving the skin dryer than before. The oils that hold the water in have usually been stripped out by the bath—especially if soap and hot water are involved. But if oil is added before the skin dries, much of the absorbed water is retained. Thus, applying a bath oil or heavy lotion directly after a bath or shower is a good method of hydrating the skin.

?

If the cells of our skin are replaced regularly, why do scars and tattoos persist indefinitely?

ANSWERED BY:

James B. Bridenstine, Department of Dermatology, University of Pittsburgh Medical Center.

O ur skin is primarily made of a protein called collagen, which is produced by cells known as fibroblasts. When the skin (or any other tissue, for that matter) is wounded, the wound-healing process initiates the generation of new fibroblasts to produce scar collagen, which is different from the collagen in normal skin. Even though individual cells within the skin periodically die and are replaced with new cells, the scar collagen remains. The only time when wounds will heal without producing scars is during the fetal stage of life, when the skin produces fetal collagen, a protein that is different from adult collagen. If we could find a way to turn on the production of fetal collagen after birth, then we could, presumably, perform scarless surgery.

Tattoos remain in the skin because the ink particles that produce the coloration are too large to be ingested by the white blood cells that patrol the body and carry foreign bodies away from the skin. The new tattoo-removing lasers work because the laser energy pulverizes the ink into microfine dust particles that are small enough to be taken in by the white blood cells and carried away.

?

Why does fat deposit on the hips and thighs of women and around the stomachs of men?

A N S W E R E D   B Y :

Patrick J. Bird, Dean, College of Health and Human Performance, University of Florida, Gainesville.

Throughout most of their lives, females have a higher percentage of body fat than males. By 25 years of age, for example, healthy-weight women have almost twice the body fat that healthy-weight men have. During the adolescent growth spurt, the rate of fat increase in girls almost doubles that of boys. It is marked by more and larger fat cells, and it is seen mostly in the gluteal-femoral area—pelvis, buttocks, and thighs—and, to a much lesser extent, in the breasts. This general acceleration in body fat accumulation, particularly sex-specific fat, is attributed mostly to changes in female hormone levels. After adolescence, the accumulation of sex-specific fat more or less stops, or decreases dramatically, in healthy-weight women, and there is usually no further increase in the number of fat cells. Fat cells in males also do not tend to multiply after adolescence.

As most women know, it is more difficult to shed fat from the pelvis, buttocks, and thighs than it is to trim down other areas of the body. During lactation, however, sex-specific fat cells are not so stubborn. They increase their fat-releasing activity and decrease their storage capacity, while at the same time fat storage increases in the mammary adipose tissue. This suggests that there is a physiological advantage to sex-specific fat. The fat stored around the pelvis, buttocks, and thighs of women appears to act as reserve storage for the energy demands of lactation. This would seem to be particularly true for habitually undernourished females.

Men tend to store excess fat in the visceral, or abdominal, region. This deposit has no apparent physiological advantage. On the contrary, it is downright dangerous. A large potbelly, where waist girth begins to exceed hip girth, is strongly associated with an increased risk of coronary artery disease, diabetes, elevated triglycerides, hypertension, cancer, and general overall mortality. Potbellies pose these health risks because the fat that produces them is metabolically more active. Abdominal fat simply breaks down more easily and enters the chemical processes related to disease quicker than sex-specific fat or fat located in other parts of the body. Unfortunately, belly fat is typically restocked as fast, or faster, than it is depleted. Body fat is, of course, necessary for life. Besides being a source of energy, it is a storage site for some vitamins, a major ingredient in brain tissue, and a structural component of all cell membranes. Moreover, it provides a padding to protect internal organs and insulates the body against the cold. But as we age, most of us tend to gain fat and weight—about 10 percent of our body weight per decade during adulthood. This stems partly from a steady decline in the metabolic rate but mostly from a decrease in physical activity. Still, getting too fat (more than 30 percent body fat in females and 25 percent in males) is associated with increased risk of disease and premature death, regardless of where the fat is stored in the body. As a society, we are severely stressing the scales to the point that obesity is now a national health epidemic.

The Dr. Is In
health and medicine

?

Why do hangovers occur?

ANSWERED BY:
Sant P. Singh, Professor and Chief of Endocrinology, Diabetes, and Metabolism, Chicago Medical School.

The alcohol hangover has been known since Biblical times: "Woe unto them that rise up early in the morning, that they may follow strong drink" (Isaiah 5:11).

Approximately 75 percent of those who drink alcohol to intoxication will experience a hangover. Consumption of relatively large amounts of alcohol leads to more severe symptoms, which include headache, nausea, vomiting, thirst and dryness of mouth, tremors, dizziness, fatigue, and muscle cramps. Often there is an accompanying slump in occupational, cognitive, or visual-spatial skills. Other symptoms, such as tachycardia (rapid heartbeat) and changes in blood pressure, might go unnoticed by the sufferer.

Although still under debate, the cause and mechanism of a hangover seem to involve several factors. Hangover has been suggested to be an early stage of alcohol withdrawal. Acetaldehyde, a breakdown product of alcohol metabolism, plays a role in producing hangover symptoms. Chemicals formed during alcohol processing and maturation known as congeners increase the frequency and severity of a hangover. Liquors such as brandy, wine, tequila, whiskey, and other dark liquors containing congeners tend to produce severe hangovers, whereas clear liquors (such as white rum, vodka, and gin) cause hangovers less frequently. Researchers have shown that severe hangovers occurred in 33 percent of subjects who ingested bourbon (which is high in congeners) but in only 3 percent of those who consumed the same dose of vodka (which is low in congeners). As a rule of thumb, the darker a liquor's color, the more congeners it contains.

People with hangovers show changes in the blood levels of several hormones, which are often responsible for some of the hangover symptoms. For example, alcohol inhibits antidiuretic hormone, which leads to excessive urination and dehydration. Dehydration accentuates the symptoms of a hangover. Other factors that contribute to an alcohol hangover include consumption of larger quantities of alcohol than the person can tolerate. Individuals who drink alcohol rapidly, or without food, or without diluting it with nonalcoholic beverages are more prone to developing a hangover. Mixing different alcoholic drinks can also cause a hangover. Additionally, smoking, loud music, flashing lights, and decreased quality and quantity of sleep can exacerbate hangover headaches.

One can diminish the severity of a hangover by paying attention to the amount and type of alcohol consumed, as well as controlling the other factors mentioned above. It is not clear that sugar-containing foods ease hangover symptoms, but sugar and fluids can help overcome hypoglycemia and dehydration, and antacids can help alleviate nausea. To reduce headache, anti-inflammatory drugs should be used cautiously.

?

Why does reading in a moving car cause motion sickness?

ANSWERED BY:

Timothy C. Hain, Professor of Neurology, Otolaryngology, and Physical Therapy/Human Movement Science, Northwestern University Medical School, Chicago, Illinois; and Charles M. Oman, Director, Man Vehicle Laboratory, Massachusetts Institute of Technology, Cambridge, Massachusetts, and Leader, Neurovestibular Research Program, NASA National Space Biomedical Institute.

In order for a person to estimate his location, the brain combines information from a variety of sources, including sight, touch, joint position, the inner ear, and its own expectations. The inner ear is particularly important because it contains sensors for both angular motion and linear motion. Under most circumstances, the senses and expectations all agree. It is when they disagree that motion sickness occurs. Motion sickness usually combines elements of spatial disorientation, nausea, and vomiting.

Consider the situation when one is reading in the backseat of a car. Your eyes, fixed on the book with the peripheral vision seeing the interior of the car, say that you are still. But as the car goes over bumps, turns, or changes its velocity, your ears disagree. This is why motion sickness is common in this situation. If you have this sort of reaction it is usually helpful to stop reading and look out the window. The driver of the car is generally least likely to suffer from motion sickness, because he not only has accurate sensory information from his ears, eyes, and touch, but he is also controlling the car and can therefore anticipate turns, accelerations, and decelerations. This position allows him to better calibrate his expectations of movement with the car's actual movement.

?

Why do we get the flu more often in the winter than in other seasons?

ANSWERED BY:
Hugh D. Niall, Medical Doctor and Chief Executive Officer, Biota Holdings Limited, Melbourne, Australia.

Every winter in the United States and other countries with largely temperate climates, there is a sharp rise in the incidence of respiratory infections, the milder of which are popularly described as "colds" and the more severe as "flu." These are caused by quite different viruses, but the distinction is blurred by an understandable tendency of some people who have colds to exaggerate the severity of their illness and lay claim to the status of being a victim of influenza.

This means that true "flu" is really a less common but a much more severe illness than many people realize. It nevertheless infects about 10 percent of the population each year. This percentage can rise to 25 or 30 percent in an epidemic year. For comparison, adults in the United States average two to four colds per year and children six to eight.

Flu is characterized by the quite sudden onset of feverishness, with a sore throat and nasal discharge, chills, headaches, muscle aches, and a loss of appetite, usually with a fever of 100 to 104 degrees Fahrenheit. Over the next few days, the general symptoms may improve but the local symptoms (sore throat, cough) get worse. In an uncomplicated case, the patient will be much improved after five to seven days but may take up to two weeks or even longer to recover completely. Flu can lead to serious complications, including bronchitis, viral or bacterial pneumonia, and even death in elderly and chronically ill patients. Twenty thousand or more people die of flu in the United States each year.

Why flu is a winter disease is not fully known. However, flu is spread largely by droplet (aerosol) infection from individuals with a high viral level in their nasal and throat secretions, sneezing and coughing on anyone close at hand. The aerosol droplets of the "right" size (thought to be about 1.5 micrometers in diameter) remain airborne and are breathed into the nose or lungs of the next victim. Situations in which people are crowded together are more common in cold or wet weather—and so perhaps this contributes to spreading the flu at these times. It is interesting that in equatorial countries, flu occurs throughout the year, but it is highest in the monsoon or rainy season.

Several recent developments promise to increase our understanding of flu. There are now drugs for influenza (neuraminidase inhibitors) that will potentially treat all strains of this virus; and new tests in development will provide an on-the-spot diagnosis in 15 minutes or less.

These advances should lead to flu being accurately diagnosed and treated.

?

What happens when you get a sunburn?

ANSWERED BY:

Jeffrey M. Sobell, Assistant Professor of Dermatology and Director of Photomedicine, Tufts University School of Medicine, Medford, Massachusetts.

A sunburn—manifested by cutaneous redness, swelling, and pain—is a toxic reaction caused by exposure to the sun's ultraviolet radiation. Although the precise mechanism by which a sunburn occurs has not been clearly identified, complex chemical reactions and pathways take place that most likely result in the well-known clinical symptoms.

The energy from ultraviolet radiation can damage molecules in the skin, most importantly DNA. One consequence of this is the synthesis of different proteins and enzymes. The effects of these proteins lead to dilation of the cutaneous blood vessels and recruitment of inflammatory cells. This, in turn, produces a sunburn's characteristic redness, swelling, and pain. Once the signal of excessive radiation exposure is initiated, it generally takes four to six hours for these proteins to generate. Sunburn symptoms thus don't appear until well after exposure. (DNA damage can also result in the destruction of the involved skin cell. This is one of the reasons why skin peels after a bad sunburn.)

The body does have mechanisms to repair damaged DNA after ultraviolet exposure. But as the frequency of sunlight exposure increases, so, too, does the probability that some of that damage will escape repair. This mutated DNA may eventually lead to skin cancer.

The warmth of a sunburn generally stems from increased blood flow to the exposed site. I am unaware of any temperature measurements of sunburned skin, but I suspect that even though the burned skin seems much warmer, it would still be close to 98.6 degrees Fahrenheit. Any slight elevation in temperature would be a result of the inflammatory response generated from the chemical processes induced by ultraviolet radiation.

?

There are many kinds of cancer, so why is there no heart cancer?

ANSWERED BY:

Alex Aller, Manager of Cell Biology and Immunology Research, Cancer Therapeutics Department, Southern Research Institute, Birmingham, Alabama.

A ny cell in the body has the potential to become malignant; thus cancer can, in fact, affect the heart. Cancer arises from mutations in the DNA of a cell. Usually a cancerous cell undergoes several mutations before it becomes a deadly, invasive cancer. Most of these mutations occur when the cell is dividing and replicating its DNA. The only way for a cell to propagate a mutation is to divide and pass those mutations on to its daughter cells. With regard to the heart cells, however, they just go right on pumping and doing their job and don't replicate to make new heart cells unless there has been some injury. With so little cell division going on in the heart, there is very little chance for mutations to occur and get passed on to daughter cells.

Now think about the types of cancer that are most common—breast, colon, and skin, among others. Most of the cells in these tissues are replacing themselves all the time. Breast tissue is constantly affected by hormones and is always growing and shrinking. The lining of the colon is continually sloughing off and being replaced. The same is true of the skin. In addition, skin and colon cells are constantly being exposed to things that induce mutations—ultraviolet rays for the skin and carcinogens in food for the colon. As a result, the likelihood of mutations is higher, and there is ample opportunity to pass these mutations on to daughter cells during cell division. This is why these types of cancer are common. The heart, in contrast, doesn't get exposed to many carcinogens, just those in the blood. That, combined with the fact that the heart cells do not replicate often, is why you don't see much cancer of the heart muscle. Indeed, according to cancer statistics, it does not appear to occur at any measurable rate.

?

Is there any proof that Alzheimer's disease is related to exposure to aluminum—for instance, by using aluminum frying pans?

ANSWERED BY:
Zaven S. Khachaturian, Director of the Ronald and Nancy Reagan Research Institute, Chicago, Illinois.

This issue has been the subject of many studies, workshops, and reports since the early 1970s. Unfortunately, there is no clear-cut answer either to implicate or to absolve the role of aluminum in causing Alzheimer's disease. It is not clear whether the aluminum found in the brain of an Alzheimer's victim got there because there is disease already in progress or if the aluminum starts the process.

In the mind of many scientists, if aluminum does play a role, it is most likely a secondary one. This reasoning is based on the fact that aluminum is one of the most abundant and pervasive elements. It is found everywhere: It is in the water we drink, it is in the dust we breathe, it is in many of the substances we use everyday such as soda in glass bottles, food preservatives, many cosmetics, and food dyes. Even if we stop using pots and pans or underarm deodorants, it will be virtually impossible to avoid aluminum. Given this type of exposure of the general population, if aluminum is playing a major role, then one would expect the numbers of people affected by Alzheimer's to be much higher than they are found in epidemiological studies.

?

How long can the average person survive without water?

ANSWERED BY:
Randall K. Packer, Professor of Biology,
George Washington University, Washington, D.C.

An adult in comfortable surroundings can survive for a week or more with no, or very limited, water intake. Under the most extreme conditions, however, death can come rather quickly. For example, a child left in a hot car or an athlete exercising hard in hot weather can dehydrate, overheat, and die in a period of a few hours.

To stay healthy, humans must maintain water balance, which means that water losses must be made up for by water intake. We get water from food and drink and lose it as sweat and urine (a small amount is also present in feces). Another major route of water loss usually goes unnoticed: Because we exhale air that is water saturated, we lose water each time we exhale. On a cold day we see this water in the air as it condenses.

Exposure to a hot environment and vigorous exercise both increase body temperature. The only physiological mechanism humans have to keep from overheating is sweating. Evaporation of sweat cools blood in the vessels in the skin, which helps to cool the entire body. Under extreme conditions, an adult can lose between 1 and 1.5 liters of sweat an hour. If that lost water is not replaced, the total volume of body fluid can fall quickly and, most dangerously, blood volume may drop. If this happens, two potentially life-threatening problems arise: Sweating stops and body temperature can soar even higher, while blood pressure decreases because of the low blood volume. Under such conditions, death occurs quickly. Because of their relatively larger skin surface-to-volume ratio, children are especially susceptible to rapid overheating and dehydration.

4

As a Matter of Fact

Elementary, My Dear Watson . . .
the elements

?

Why doesn't stainless steel rust?

ANSWERED BY:
Michael L. Free, Metallurgical Engineer,
University of Utah, Salt Lake City.

Stainless steel remains stainless, or does not rust, because of the interaction between its alloying elements and the environment. Stainless steel contains iron, chromium, manganese, silicon, carbon, and, in many cases, significant amounts of nickel and molybdenum. These elements react with oxygen from water and air to form a very thin, stable film that consists of such corrosion products as metal oxides and hydroxides. Chromium plays a dominant role in reacting with oxygen to form this corrosion product film. In fact, all stainless steels by definition contain at least 10 percent chromium.

The presence of the stable film prevents additional corrosion by acting as a barrier that limits oxygen and water access to the underlying metal surface. Because the film forms so readily and tightly, even only a few atomic layers reduce the rate of corrosion to very low levels. The fact that the film is much thinner than the wavelength of light makes it difficult to see without the aid of modern instruments. Thus, although the steel is corroded on the atomic level, it appears stainless. Common inexpensive steel, in contrast, reacts with oxygen from water to form a relatively unstable iron oxide/hydroxide film that continues to grow with time and exposure to water and air. As such, this film, otherwise known as rust, achieves sufficient thickness to make it easily observable soon after exposure to water and air.

?

If nothing sticks to Teflon, how does it stick to pans?

ANSWERED BY:

Andrew J. Lovinger, Director, Polymers Program, National Science Foundation, Arlington, Virginia.

Teflon is a trademark of DuPont for a plastic material known as polytetrafluoroethylene. The secret to Teflon's slick surface lies in the fluorine atoms enveloping its molecules. These fluorine atoms repel almost all other materials, preventing them from adhering to Teflon.

We can use two techniques to make Teflon itself stick to surfaces of items such as pots and pans. The first is "sintering," a process similar to melting, in which the Teflon is heated at a very high temperature and pressed firmly onto a surface. When the material cools down to room temperature, however, chances are it will eventually peel away. Chemically modifying the side of the Teflon that you want to have "stick" yields better results. By bombarding it with ions in a high vacuum under an electric field, or "plasma," we can break away many of the fluorine atoms on the surface that we want to make sticky.

We can then substitute other groups, such as oxygen, that adhere strongly to surfaces.

Though perhaps best known as a cookware coating, Teflon has a wide range of applications, from insulating data communications cables to repelling water and stains from clothing and upholstery.

?

What determines whether a substance is transparent?

ANSWERED BY:

Morton Tavel, Professor of Physics, Vassar College, Poughkeepsie, New York.

The propagation of light through a solid is a complex process that involves not just the passage of the incident light but also reradiation of that light by the electronic structure of the solid. Simply stated, a solid material will appear transparent if there are no processes that compete with transmission, either by absorbing the light or by scattering it in other directions. In pure silicon, there is a very strong absorptive process at work: The incident visible light is absorbed by electrons that then move from one electron energy state to another. Glass, being silicon dioxide—not pure silicon—does not have this band structure, so it cannot absorb light as pure silicon does. Sand, on the other hand, is also silicon dioxide, but it is so filled with impurities that light simply scatters outward incoherently and does not pass through to a noticeable extent.

The electronic structure of solids also explains why metals are shiny. Pure metals reflect light but do not transmit it, because they are filled with free electrons. These electrons reradiate the light in the direction opposite from which it arrived (reflection), but they interfere with the light that would proceed in the forward direction, preventing transmission.

If You Can't Stand the Heat, Get Out of the Kitchen!
everyday chemistry

?

Why do my eyes tear when I peel an onion?

ANSWERED BY:
Thomas Scott, Dean, College of Sciences, San Diego State University.

In this case, tears are the price we pay for flavor and nutritional benefits. The rowdy onion joins the aristocratic shallot, gentle leek, herbaceous chive, sharp scallion, and assertive garlic among the 500 species of the genus allium. Allium cepa is an ancient vegetable, known to Alexander the Great and eaten by the Israelites during their Egyptian bondage. Indeed, his charges chastened Moses for leading them away from the onions and other flavorful foods that they had come to relish while in captivity. And with good reason: The onion is a rich source of nutrients (such as vitamins B, C, and G), protein, starch, and other essential compounds. The chemicals in onions are effective agents against fungal and bacterial growth; they protect against stomach, colon, and skin cancers; they have anti-inflammatory, antiallergenic, antiasthmatic, and antidiabetic properties; they treat causes of cardiovascular disorders, including hypertension, hyperglycemia, and hyperlipidemia; and they inhibit platelet aggregation.

The tears come from the volatile oils that help to give allium vegetables their distinctive flavors and that contain a class of organic molecules known as amino acid sulfoxides. Slicing an onion's tissue releases enzymes called allinases, which convert these molecules to sulfenic acids. These acids, in turn, rearrange to form syn-propanethial-S-oxide, which triggers the tears. They also condense to form thiosulfinates, the cause of the pungent odor associated with chopping onions and often mistakenly blamed for the weepy eye. The formation of syn-propanethial-S-oxide peaks about 30 seconds after the onion is first peeled and completes its cycle of chemical evolution over about five minutes.

The effects on the eye are all too familiar: a burning sensation and tears. The eye's protective front surface, the cornea, is densely populated with sensory fibers of the ciliary nerve, a branch of the massive trigeminal nerve that brings touch, temperature, and pain sensations from the face and the front of the head to the brain. The cornea also has a smaller number of autonomic motor fibers that activate the lachrymal (tear) glands. Free nerve endings detect syn-propanethial-S-oxide on the cornea and drive activity in the ciliary nerve, which the central nervous system registers as a burning sensation. This nerve activity reflexively activates the autonomic fibers, which then carry a signal back to the eye to order the lachrymal glands to wash the irritant away.

There are several solutions to the problem of onion tears. You can heat onions before chopping to denature the enzymes. You might also try to limit contact with the vapors: Chop onions on a breezy porch, under a steady stream of water, or mechanically in a closed container. Some say that wearing contact lenses helps.

?

Why do spicy (or "hot") foods cause the same physical reactions as heat?

ANSWERED BY:
Barry Green, John B. Pierce Laboratory, New Haven, Connecticut.

The answer hinges on the fact that spicy foods excite the receptors in the skin that normally respond to heat. Those receptors are pain fibers, technically known as polymodal nociceptors. They respond to temperature extremes and to intense mechanical stimulation, such as pinching and cutting; they also respond to certain chemical influences. The central nervous system can be confused or fooled when these pain fibers are stimulated by a chemical, like that in chile peppers, which triggers an ambiguous neural response.

So how does the brain decide whether the mouth is being pinched, cut, burned, or affected by chemicals? Scientists are not certain how the process works, but probably the brain makes a judgment based on the type and variety of stimuli being received. Stimulus to the nociceptors alone might indicate dangerous, extreme temperature. But capsaicin, the active ingredient in chile peppers, also stimulates the nerves that respond only to mild increases in temperature—the ones that give the sensation of moderate warmth. So capsaicin sends two messages to the brain: "I am an intense stimulus," and "I am warmth." Together these stimuli define the sensation of a burn, rather than a pinch or a cut.

The central nervous system reacts to whatever the sensory system tells it is going on. Therefore, the pattern of activity from pain and warm nerve fibers triggers both, the sensations and the physical reactions of heat, including the dilation of blood vessels, sweating, and flushing.

Most people think of the "burn" of spicy food as a form of taste. In fact, the two sensory experiences are related but are very distinct. They innervate the tongue the same way, but the pain system that is triggered by capsaicin is everywhere on the body, so one can get thermal effects everywhere. Some liniments contain compounds that produce similar temperature stimuli to the nerves in the skin. Menthol acts in much the same way as capsaicin, but in this case, it stimulates the fibers that register cold temperatures, not those that respond to warmth. This is why products containing menthol have names like "Icy Hot"—menthol stimulates both the hot (pain) and cold receptors, sending the brain a really ambiguous signal. That difference explains why there is no confusing menthol and capsaicin: One gives rise to a cool burn, the other to a hot burn.

The sensations produced by menthol and capsaicin are accidents of human physiology—we obviously did not evolve receptors to react to these compounds. The chemicals fool pain receptors whose real purpose is to register critical events, like damage to the skin and the inflammation that often results. The tenderness around an injury is caused in part by the response of these same nerves to chemicals released in the skin. We humans are peculiar creatures—we've taken a nerve response that normally signals danger and turned it into something pleasurable.

?

Why does bruised fruit turn brown?

ANSWERED BY:
Jonathan H. Gutow, Department of Chemistry, University of Wisconsin at Oshkosh.

The short answer is that chemical compounds in the fruit are oxidized when the skin of the fruit—and hence the walls and membranes of the cells within the fruit—is ruptured, allowing oxygen in. These compounds react with oxygen, usually incorporating it into their molecular structure. Many of the resulting oxidized organic compounds have a brown color. Citric acid is very easily oxidized and can be used to scavenge oxygen to keep the fruit from turning brown. This is why, for instance, sliced apples will remain white for a much longer time if they are first dipped in lemon juice.

?

How is caffeine removed to produce decaffeinated coffee?

ANSWERED BY:
Fergus M. Clydesdale, Head of the Food Science Department, University of Massachusetts at Amherst.

There are currently three processes, all of which begin with moistening the green or roasted beans to make the caffeine soluble. In the first method, called water processing, the moistened coffee beans are soaked in a mixture of water and green-coffee extract that has previously been caffeine-reduced. Osmosis draws the caffeine from the highly caffeine-concentrated beans into the less caffeine-concentrated solution. Afterward, the decaffeinated beans are rinsed and dried. The extracted caffeine-rich solution is passed through a bed of charcoal that has been pretreated with a carbohydrate. The carbohydrate blocks sites in the charcoal that would otherwise absorb sugars and additional compounds that contribute to the coffee's flavor but permits the absorption of caffeine. The caffeine-reduced solution, which still contains compounds that augment the taste and aroma, can then be infused into the beans. The water process is natural, in that it does not employ any harmful chemicals, but it is not very specific for caffeine, extracting some noncaffeine solids and reducing flavor. It eliminates 94 to 96 percent of the caffeine.

An alternative method extracts caffeine with a chemical solvent. The liquid solvent is circulated through a bed of moist, green coffee beans, removing the caffeine. The solvent is recaptured in an evaporator, and the beans are washed with water. Finally, the beans are steamed to remove chemical residues. Solvents, such as methylene chloride, are more specific for caffeine than charcoal is, extracting 96 to 97 percent of the caffeine and leaving behind nearly all the noncaffeine solids.

In the third approach, carbon dioxide is circulated through the beans in drums operating at roughly 250 to 300 times atmospheric pressure. At these pressures, carbon dioxide takes on unique supercritical properties, having a density similar to that of a liquid but with the diffusivity of a gas, allowing it to penetrate the beans and dissolve the caffeine. These attributes also significantly lower the pumping costs for carbon dioxide. The caffeine-rich carbon dioxide exiting the extraction vessel is channeled through charcoal or water to absorb the caffeine and is then returned to the extraction vessel. Carbon dioxide is popular because it has a relatively low pressure critical point, it is nontoxic, and it is naturally abundant. This method of decaffeination is more expensive, but it extracts 96 to 98 percent of the caffeine.

What is the difference between artificial and natural flavors?

ANSWERED BY:

Gary Reineccius, Professor, Department of Food Science and Nutrition, University of Minnesota, Minneapolis.

There is little substantive difference in the chemical compositions of natural and artificial flavorings. They are both made in a laboratory by a trained professional, a "flavorist," who blends appropriate chemicals together in the right proportions. The flavorist uses "natural" chemicals to make natural flavorings and "synthetic" chemicals to make artificial flavorings. The flavorist creating an artificial flavoring must use the *same* chemicals in his formulation as would be used to make a natural flavoring, however. Otherwise, the flavoring will not have the desired flavor. The distinction in flavorings—natural versus artificial—comes from the *source* of these identical chemicals and may be likened to saying that an apple sold in a gas station is artificial and one sold from a fruit stand is natural.

This issue is somewhat confusing to the average consumer in part because of other seeming parallels in the world. One can, for example, make a blue dye out of blueberry extract or synthetic pigments. These dyes are very different in chemical composition, yet both yield a blue color. Similarly, consider one shirt made from wool and another from nylon. Both are shirts, but they have very different chemical compositions. This diversity of building blocks is not possible in flavorings—one makes a given flavor only by using specific chemicals. Thus, if a consumer purchases an apple beverage that contains an artificial flavor, she will ingest the same primary chemicals that she would take in if she had chosen a naturally flavored apple beverage.

Artificial flavorings are simpler in composition and potentially safer because only safety-tested components are utilized. Another difference is cost. The search for "natural" sources of chemicals often requires that a manufacturer go to great lengths to obtain a given chemical. Natural coconut flavorings, for example, depend on a chemical called massoya lactone. Massoya lactone comes from the bark of the massoya tree, which grows in Malaysia. Collecting this natural chemical kills the tree because harvesters must remove the bark and extract it to obtain the lactone. Furthermore, the process is costly. This pure natural chemical is identical to the version made in an organic chemist's laboratory, yet it is much more expensive than the synthetic alternative. Consumers pay a lot for natural flavorings. But these are in fact no better in quality, nor are they safer, than their cost-effective artificial counterparts.

?

How can an artificial sweetener contain no calories?

ANSWERED BY:
Arno F. Spatola, Professor of Chemistry and Director, Institute for Molecular Diversity and Drug Design, University of Louisville, Kentucky.

Sweetness is a taste sensation that requires interaction with receptors on the tongue. Many sugar substitutes, such as saccharin and acesulfame K, also known as sunette, do not provide any calories. This means that they are not metabolized as part of the normal biochemical process that yields energy in the form of adenosine triphosphate, or ATP. In some cases, small quantities of additives such as lactose are present to improve the flow characteristics or to give bulk to a product, but the amounts are so small that they do not represent a significant source of energy.

The sugar substitute aspartame, also called NutraSweet, is more interesting. This synthetic compound is a dipeptide, composed of the two amino acids phenylalanine and aspartic acid. As with most proteins, which are chains of amino acids, it can be metabolized and used as an energy source. In general, we obtain energy in the amount of four calories (more correctly termed kilocalories) per gram of protein. This is the same value as the number of calories acquired from sugars or starches. (In contrast, each gram of fat consumed provides more than twice that amount, or about nine calories per gram.)

So if aspartame has the same number of calories per gram as common table sugar (sucrose), how is it a low-calorie sweetener? The answer is that aspartame is 160 times as sweet as sugar. That is, a single teaspoon of aspartame (4 calories) will yield the same sweetening effect as 160 teaspoons of sugar (640 calories). If 3,500 extra calories is equivalent to a gain of one pound in weight, it is easy to see why so many people turn to artificially sweetened beverages in an effort to maintain some control over their amount of body fat. But does that actually lead to weight loss? Perhaps not. Either by a physical effect, or perhaps a psychological one, many of us seem to make up the loss of sugar calories by eating or drinking other foods. For this reason, artificially sweetened diet drinks alone are hardly likely to have much of an effect on the problem of obesity in the United States.

?

Do vitamins in pills differ from those in food?

ANSWERED BY:

Christine Rosenbloom, Professor of Nutrition, Georgia State University, Atlanta, Georgia, and Spokesperson, American Dietetic Association.

Vitamins and minerals in supplements are synthetic forms of the nutrients. The word "synthetic" doesn't necessarily mean inferior, however. Even those supplements that claim to have "natural" ingredients contain some synthetic ingredients. Indeed, if a pill contained only natural ingredients, it would be the size of a golf ball. For the most part, our bodies appear to absorb synthetic forms as well as they do natural forms. The one exception seems to be vitamin E, which is better absorbed in natural form than in synthetic form. But most supplements now contain more natural vitamin E, so it is well absorbed in pill form.

For absorption to occur, a pill must dissolve and disintegrate. So when shopping for supplements, look for the USP symbol. This symbol indicates that the U.S. Pharmacopeia, an independent testing organization, has tested the supplement to make sure it will dissolve in your stomach. The absorption of nutrients in pill form is not well studied, but if they dissolve in the stomach, they are probably well absorbed.

Look for a supplement that contains about 100 percent of the daily values for nutrients. Don't spend extra money on products that are marked "high potency," "stress formula," or "laboratory approved." The supplement industry is not well regulated and claims can be made without much scientific proof. And don't forget, "food first." Foods contain substances other than vitamins and minerals for good health. Fruits, vegetables, and whole grains contain phytochemicals, or plant chemicals, that can help to fight the development and progression of many chronic diseases, including cancer.

Where There's Smoke, There's a Fire
more chemistry

?

How does a flame behave in zero gravity?

ANSWERED BY:
Kenneth D. Schlecht, Chemist, State University of New York College at Brockport.

A typical flame, such as that from a candle, produces light, heat, carbon dioxide, and water vapor. The heat causes these combustion products to expand, which lowers their density, and they rise due to buoyancy. Fresh, oxygen-containing air can thus get into the flame, further fueling the combustion process.

Because gravity is necessary for density differences to arise, neither buoyancy nor convection occur in a zero-gravity environment such as space. Consequently, the combustion products accumulate around the flame, preventing sufficient oxygen from reaching it and sustaining the combustion reaction. Ultimately the flame goes out.

In the early years of the U.S. space program, tests were

conducted on unmanned missions to ascertain what would happen to a flame in a pure oxygen environment under weightless conditions. Researchers learned that flames extinguish themselves. They ran these experiments because they hoped to have an oxygen environment for manned missions, and there was concern about the possibility of a rampant fire. Unfortunately, in 1967 fire broke out in the *Apollo I* spacecraft while it was still on the ground and three astronauts were killed. The flames didn't self-extinguish because the launch pad was not a gravity-free environment.

Oxygen could still reach a flame in a gravity-free environment if someone blew the gas into the flame or let it "diffuse" in. It is the diffusion process that spreads the scent of a perfume in a room without air circulation: The perfume slowly mixes with the air to try to achieve a uniform distribution. This process, however, is too slow to sustain a flame.

?

How does fingerprint powder work?

ANSWERED BY:

Christine Craig, Forensic Scientist, Commonwealth of Virginia; and Jason Byrd, Assistant Professor of Forensic Science and Biology, Virginia Commonwealth University, Richmond, Virginia, and Chair, American Board of Forensic Entomology.

Fingerprint patterns and characteristics are formed before birth. They will remain unchanged until decomposition destroys them after death or unless the dermal layer is injured, producing a scar. Fingerprints are unique to each individual—including identical twins—and have been used for over a century for identification and crime-solving purposes.

The skin found on the fingers, palms, and soles of the feet of humans (and some primates) is known as friction skin. This skin is unique because it does not have hair follicles or oil glands, and because it is composed of ridges that are believed to be adapted for increased friction to help when handling various objects and walking. These so-called friction ridges are composed of rows of sweat pores, or eccrine glands, that constantly secrete perspiration. This perspiration—along with grease and oil transferred from other parts of the body—adheres to the friction skin and is transferred from the skin to other surfaces when contact is made with objects. The transferred outline of the friction ridges is what is known as a latent print.

Fingerprint patterns form before birth and last a lifetime. Latent prints are not readily visible to the naked eye. As a result, these "hidden" prints must be "developed" in some way to increase their visibility and contrast. The most common method of developing latent prints on non-porous objects is to physically enhance them by applying fingerprint powder. Fingerprint powder is composed of many different ingredients that can vary greatly depending on the formula used. Most black fingerprint powders contain rosin, black ferric oxide, and lampblack. Many also contain inorganic chemicals such as lead, mercury, cadmium, copper, silicon, titanium, and bismuth. Fingerprint powder is applied by brushing it onto the surface and works by mechanically adhering to the oil and moisture components of the latent print. When the powder particles adhere to the grease or moisture forming the latent prints, it causes them to become visible. The developed latent prints are then readily observable and able to be collected, preserved, and examined.

There's No Place Like Home

Everybody Talks About It . . .
weather

?

Why do clouds float when they have so much water in them?

ANSWERED BY:
Douglas Wesley, Senior Meteorologist, Cooperative Program for Operational Meteorology, Education and Training (COMET), University Corporation for Atmospheric Research, Boulder, Colorado.

C louds are composed primarily of small water droplets and, if it's cold enough, ice crystals. The vast majority of clouds you see contain droplets and/or crystals that are too small to have any appreciable fall velocity, or downward speed. So the particles continue to float with the surrounding air. For an analogy closer to the ground, think of tiny dust particles that, when viewed against a shaft of sunlight, appear to float in the air.

The speed with which any object falls is related to its mass and surface area—which is why a feather falls more slowly than a pebble of the same weight. As a tiny water droplet grows, its mass becomes more important than its shape, and the droplet falls faster. Even a large droplet having a radius of 100 microns, or about 2 hairwidths, has a fall velocity of only about 27 centimeters per second (cm/s). And because ice crystals have more irregular shapes, their fall velocities are relatively smaller.

Another way to illustrate the relative lightness of clouds is to compare the total mass of a cloud to the mass of the air in which it resides. Consider a hypothetical but typical small cloud at an altitude of 10,000 feet, comprising 1 cubic kilometer and having a liquid water content of 1 gram per cubic meter. The total mass of the cloud particles is about 1 million kilograms, which is roughly equivalent to the weight of 500 automobiles. But the total mass of the air in that same cubic kilometer is about 1 billion kilograms—1,000 times heavier than the liquid!

So, even though typical clouds do contain a lot of water, this water is spread out for miles in the form of tiny water droplets or crystals, which are so small that the effect of gravity on them is negligible. Thus, from our vantage point on the ground, clouds seem to float in the sky.

?

What causes thunder?

ANSWERED BY:
**Richard Brill, Associate Professor,
Honolulu Community College.**

Thunder is caused by lightning, which is essentially a stream of electrons flowing between or within clouds or between a cloud and the ground. The air surrounding the lighted stream of electrons becomes hot—like the burner on an electric range, only much hotter. It also moves faster than regular, unheated air and so becomes like an explosion of light and heat.

This explosion creates a tube of partial vacuum along the path of least resistance followed by the lightning through the air. As the air in this tube rapidly expands and contracts, it creates a clap not unlike the clap of two hands. This sound causes the tube to vibrate like a tubular drumhead. As the sound echoes and reverberates, it produces the rumbling we call thunder. We can hear these rumbles from a great distance, even when the lightning causing them isn't yet visible.

When the lightning is within sight, however, we see it first because the speed of sound in air is considerably slower that that of the electron flow. Thus, the sound behaves more like a shock wave than an ordinary sound wave. The shock wave follows the path of the electrons like a fist in a sock. The speed of sound is even more insignificant when compared to the speed of light. The light from the flash reaches us in a fraction of a second, whereas the sound lags along like a snail following an interplanetary rocket.

The audiovisual spectacle of thunder and lightning is a combination of the dynamics of the vibration of air molecules and their disturbance by electrical forces. It is an awesome show—and one that reminds all of us of the powers of nature and our own insignificance in relation to them.

?

Why are snowflakes symmetrical?

ANSWERED BY:

Miriam Rossi, Associate Professor of Structural Chemistry, Vassar College, Poughkeepsie, New York.

Snowflakes reflect the internal order of water molecules as they arrange themselves in their solid forms: snow and ice. As water molecules begin to freeze, they form weak hydrogen bonds with one another. The growth of snowflakes (or any substance changing from a liquid to a solid) is known as crystallization. The molecules align themselves in their lowest-energy state, which maximizes the attractive forces among them and minimizes the repulsive ones. In the water ice found on the earth, each molecule is linked by hydrogen bonds to four other molecules, creating a lattice structure.

As a result, the water molecules move into prearranged spaces. The most basic shape is a hexagonal prism, with hexagons on the top and bottom and six rectangular-shape sides. This ordering process is much like tiling a floor: Once the pattern is chosen and the first tiles are placed, all the other tiles must go in predetermined spaces to maintain the pattern. Water molecules settle themselves in low-energy locations that fit the spaces and maintain symmetry; in this way, the arms of the snowflake are made.

There are many types of snowflakes. The differentiation occurs because each snowflake forms in the atmosphere, which is complex and variable. A snow crystal may begin developing in one way and then change in response to alterations in temperature or humidity. The basic hexagonal symmetry is preserved, but the ice crystal branches off in new directions.

?

Why are some rainbows bigger than others?

ANSWERED BY:
**William Ducker, Professor of Chemistry, Virginia
Polytechnic Institute and State University, Blacksburg.**

Rainbows appear when light originating from the sun is refracted and reflected by small water droplets suspended in the air. When a water droplet refracts sunlight, it changes the angle at which the light travels both as it enters and exits the droplet. How much the angle changes depends on how the light interacts with the water molecules.

Light from the sun is a mixture of different colors that look white when they are superimposed. Each color interacts with water molecules differently. And as a result, the water droplet changes the angle of each color differently. The water droplets in the rainbow cause these different colors to be viewed at different angles and thus to appear as separate colors.

The physics of rainbow formation dictates the angle between the viewer, the sun, and the rainbow—but not the absolute position or actual size of the rainbow. A rainbow appears to move as a viewer moves (hindering the search for a pot of gold). And its size is also a matter of human perception. The only true information received by our brain is the angle. From experience, we know that the boundary of an object that is far away subtends a smaller angle than one that is near. For example, a single head can obscure your entire view of a movie screen if the head is close enough to you. The human brain uses this experience in reverse. It measures the angle subtended by the rainbow and then looks at other features, such as mountains, in the surroundings. If the other features are far away, then the brain interprets the rainbow as a very large object. Conversely, if the rainbow appears to be near the observer, it also looks much smaller.

?

What is the meaning of the phrase, "It is too
cold to snow"? Doesn't it have to be cold for
it to snow?

ANSWERED BY:

Matt Peroutka, Meteorologist, National Weather
Service's Techniques Development Laboratory,
Silver Spring, Maryland.

It has to be cold for it to snow, if your definition of cold
is like that of most folks who live in the mid-latitudes.
But the atmosphere must contain moisture to generate
snow—and very cold air contains very little moisture.
Once the air temperature at ground level drops below
about -10 degrees Fahrenheit (-20 degrees Celsius), snowfall becomes unlikely in most places. Therefore, significant
snowfall at such very low temperatures is rare.

?

Why do hurricanes hit the East Coast of the United States but never the West Coast?

ANSWERED BY:

Chris W. Landsea, Researcher, Atlantic Oceanographic and Meteorological Laboratory/Hurricane Research Division, National Oceanic and Atmospheric Administration (NOAA), Miami, Florida.

Hurricanes form both in the Atlantic basin, to the east of the continental United States (that is, in the Atlantic Ocean, the Gulf of Mexico, and the Caribbean Sea), and in the Northeast Pacific basin, to the west of the Unites States. The hurricanes in the Northeast Pacific almost never hit the United States, however, whereas the ones in the Atlantic basin strike the U.S. mainland just less than twice a year on average.

There are two main reasons for this disparity. The first is that hurricanes in the Northern Hemisphere form at tropical and subtropical latitudes and then tend to move toward the west-northwest. In the Atlantic, such a motion often brings the hurricane into the vicinity of the East Coast of the United States. In the Northeast Pacific, the same west-northwest track carries hurricanes farther off-shore, well away from the U.S. West Coast.

The second factor is the difference in water temperatures along the U.S. East and West coasts. Along the East Coast, the Gulf Stream provides a source of warm (above 80 degrees Fahrenheit, or 26.5 degrees Celsius) waters, which helps to maintain the hurricane. Along the West Coast, however, ocean-surface temperatures rarely rise above the lower 70s Fahrenheit (the low 20s Celsius), even in the middle of summer. Such relatively cool temperatures do not provide enough thermal energy to sustain a hurricane's strength. So the occasional Northeast Pacific hurricane that does track back toward the United States encounters the cooler waters of the Pacific, which can quickly reduce the storm's strength.

Up Above
the atmosphere

?

If chlorofluorocarbons are heavier than air,
how do they reach the ozone layer?

ANSWERED BY:

F. Sherwood Rowland, University of California at Irvine,
who won a Nobel Prize for his work on
atmospheric chemistry.

Although the CFC molecules are indeed several times heavier than air, thousands of measurements have been made from balloons, aircraft, and satellites demonstrating that the CFCs are actually present in the stratosphere. The atmosphere is not stagnant. Winds mix the atmosphere to altitudes far above the top of the stratosphere much faster than molecules can settle according to their weight. Gases such as CFCs that are insoluble in water and relatively unreactive in the lower atmosphere (below about 10 kilometers) are quickly mixed and therefore reach the stratosphere regardless of their weight.

?

What determines the shape of a mushroom cloud after a nuclear explosion?

ANSWERED BY:

David Dearborn, Physicist, Lawrence Livermore National Laboratory, Livermore, California.

A mushroom cloud forms when an explosion creates a very hot bubble of gas. In the case of a nuclear detonation, the bomb emits a blast of x-rays, which ionize and heat the surrounding air; that hot bubble of gas is known as a fireball. The hot air is buoyant, so it quickly rises and expands. The rising cloud creates a powerful updraft that picks up dust, forming the stem of the mushroom cloud.

The central part of the fireball is hottest, creating a rolling motion as it interacts with the outer portions. Thermal instabilities, called Kelvin-Helmholtz instabilities, occur at the interface between the fireball and the neighboring cool air. If you watch a movie of a nuclear detonation, you can see material swirling outward as a result.

All atomic bombs produce a bulge and a stem, but the really huge, flat clouds—the ones that could be described only as mushrooms—come from the explosions caused by thermonuclear weapons or hydrogen bombs. The fireball from an H-bomb rises so high that it hits the tropopause, the boundary between the troposphere and the stratosphere. There is a strong temperature gradient at the tropopause, which prevents the two layers of the atmosphere from mixing much. The hot bubble of the fireball initially expands and rises. By the time the bubble has risen from sea level to the tropopause, it is no longer hot enough to break through the boundary. At that point, the fireball flattens out; it can no longer expand upward, so it expands to the side into an exaggerated mushroom cap. The same thing happens to big summer thunderclouds when they rise up to the tropopause, producing a characteristic flattened-anvil shape.

The Upper Crust
earth's surface and below

?

How do volcanoes affect world climate?

ANSWERED BY:
Karen Harpp, Assistant Professor of Geology, Colgate University, Hamilton, New York.

I n 1784, Benjamin Franklin made what may have been the first connection between volcanoes and global climate while stationed in Paris as the first diplomatic representative of the United States of America. He observed that during the summer of 1783, the climate was abnormally cold, both in Europe and back in the United States. The ground froze early, the first snow stayed on the ground without melting, the winter was more severe than usual, and there seemed to be "a constant fog over all Europe, and a great part of North America."

What Franklin observed was indeed the result of volcanic activity. An enormous eruption of the Laki fissure system (a chain of volcanoes in which the lava erupts through a crack in the ground instead of from a single point) in Iceland caused the disruptions. The Laki eruptions produced about 14 cubic kilometers of basalt (thin, black, fluid lava) during more than eight months of activity. More important in terms of global climate, however, the Laki event also produced an ash cloud that may have reached up into the stratosphere. This cloud caused a dense haze across Europe that dimmed the sun, perhaps as far west as Siberia. In addition to ash, the cloud consisted primarily of vast quantities of sulfur dioxide (SO_2), hydrogen chloride ($HC1$), and hydrogen fluoride (HF) gases. The gases combined with water in the atmosphere to produce acid rain, destroying crops and killing livestock. The effects, of course, were most severe in Iceland; ultimately, more than 75 percent of Iceland's livestock and 25 percent of its human population died from famine or the toxic impact of the Laki eruption clouds. Consequences were also felt far beyond Iceland. Temperature data from the United States indicate that record lows occurred during the winter of 1783–1784. In fact, the temperature decreased about 1 degree Celsius in the Northern Hemisphere overall. That may not sound like much, but it had enormous effects in terms of food supplies and the survival of people across the Northern Hemisphere. For comparison, the global temperature of the most recent Ice Age was only about 5 degrees Celsius below the current average.

There are many reasons that large volcanic eruptions have such far-reaching effects on global climate. The most substantive climatic effect results from the production of atmospheric haze. Large eruptions inject ash particles and sulfur-rich gases into the troposphere and stratosphere, and these clouds can circle the globe within weeks of the volcanic activity. The small ash particles decrease the amount of sunlight reaching the surface of the earth and lower average global temperatures. The sulfurous gases combine with water in the atmosphere to form acidic aerosols that also absorb incoming solar radiation and scatter it back out into space.

Initially, scientists believed that it was volcanoes' stratospheric ash clouds that had the dominant effect on global temperatures. The 1982 eruption of El Chichón in Mexico, however, altered that view. Only two years earlier, the major Mount St. Helens eruption had lowered global temperatures by about 0.1 degree Celsius. The much smaller eruption of El Chichón, in contrast, had three to five times the global cooling effect worldwide. Despite its smaller ash cloud, El Chichón emitted more than 40 times the volume of sulfur-rich gases produced by Mount St. Helens. Sulfate aerosols appear to take several years to settle out of the atmosphere, which is one of the reasons their effects are so widespread and long lasting.

The atmospheric effects of volcanic eruptions were confirmed by the 1991 eruption of Mount Pinatubo in the Philippines. Pinatubo's eruption cloud reached over 40 kilometers into the atmosphere and ejected about 17 million tons of SO_2, just over two times that of El Chichón in 1982. The sulfur-rich aerosols circled the globe within three weeks and produced a global cooling effect approximately twice that of El Chichón. The Northern Hemisphere cooled by up to 0.6 degree Celsius during 1992 and 1993. Moreover, the aerosol particles may have contributed to an accelerated rate of ozone depletion during that same period. Interestingly, some scientists argue that without the cooling effect of major volcanic eruptions such as El Chichón and Mount Pinatubo, global warming effects caused by human activities would have been far more substantial.

Major volcanic eruptions have additional climatic effects beyond global temperature decreases and acid rain. Ash and aerosol particles suspended in the atmosphere scatter light of red wavelengths, often resulting in brilliantly colored sunsets and sunrises around the world. The spectacular optical effects of the 1883 Krakatau eruption cloud were observed across the globe, and they may have inspired numerous artists and writers in their work. The luminous, vibrant renderings of the fiery late day skyline above the Thames River in London by the British painter William Ascroft, for instance, may be the result of the distant Krakatau eruption.

?

Where do geysers get their water from?

ANSWERED BY:
**Karen Harpp, Assistant Professor of Geology,
Colgate University, Hamilton, New York.**

A geyser eruption is one of nature's most impressive displays of hydrothermal energy. They occur where *magma* lies just below the earth's surface, particularly in volcanic regions such as Iceland or New Zealand, and in places that have been volcanically active in the past, including Yellowstone in Wyoming. Water from rain or melted snow percolates into the ground through cracks and fractures, and it interacts with the hot underlying rocks. The water reaches temperatures far above where it would boil on the earth's surface (about 100 degrees Celsius), but because there is so much rock above the water (sometimes up to several miles), the water does not boil. Instead it becomes superheated and pressurized. Once enough pressure builds up, the superheated water will overcome the weight of the overlying rocks and burst out of the ground in an explosive steam eruption. It basically works like a teapot with a closed lid; only when enough pressure builds up from accumulating hot water and steam is there enough force for the steam to burst out through the top and activate the whistle.

One of the most fascinating aspects of geysers is that once they form, they become self-perpetuating. After the initial eruption of hot water, the pressure on the superheated groundwater is reduced, which causes some of it to flash to steam (when you reduce pressure on a liquid, it becomes easier for the individual molecules to escape into the vapor phase—even without an increase in temperature). Because it is a gas, the steam expands rapidly, causing it to burst upward through the small, tight fractures in the rock, forcing out any hot water that was left behind after the initial blast. Once the channels are empty, the eruption ends and the cycle begins anew. More water seeps into the hot areas along the fractures, heats up, and starts building pressure all over again.

Some geysers go through the pressure-building cycle

quickly, producing fountains every few minutes. Old Faithful in Yellowstone—one of the most famous geysers in the world—puts on its show approximately every 80 minutes and can reach up to nearly 200 feet in height. If you observe Old Faithful in action for several eruptions, you can begin to see how the pressure-building process plays a role in its behavior. When the eruption is short (less than about 2 minutes), the next blast usually happens within about 45 minutes, a relatively short interval of inactivity for this geyser. But when the eruption is more powerful (up to 5 minutes long), it will take more time for the pressure to reach the critical level, and you could be waiting nearly an hour and a half for the next event.

The name "geyser" comes from the Icelandic word "geysir," for "gush," and the original namesake is found in Iceland, in the Haukadalur Basin. Unfortunately, it only erupts occasionally, which illustrates another important point about geysers. They are ephemeral features, usually only lasting for several thousand years or until a major geologic event occurs. For instance, an earthquake of magnitude 6.1 in 1975 in the Norris Geyser Basin of Yellowstone caused greater water volumes to flow in some geysers and decreased the activity of others by rearranging groundwater paths and fracture patterns throughout the area. Even earthquakes as small as magnitude 4 can change the behavior and eruption intervals of geysers.

Moreover, it turns out that earthquakes may be necessary to keep some geysers active. Over time, water conduits feeding geysers can become narrower or even clog as a result of the deposition of minerals from the hot water (just like pipes can clog in areas with hard, mineral-rich water). Frequent tremors may break up the deposits or create new fractures, providing alternative pathways for the water to move through the rock and keep the geysers supplied.

?

How do scientists measure the temperature of the earth's core?

ANSWERED BY:

Gregory Lyzenga, Associate Professor of Physics, Harvey Mudd College, Claremont, California.

The center of the earth lies 6,400 kilometers (4,000 miles) beneath our feet, but the deepest that it has ever been possible to drill to make direct measurements of temperature (or other physical quantities) is just about 10 kilometers (6 miles). As a result, scientists must infer the temperature in the earth's deep interior indirectly. Observing the speed at which seismic waves pass through the earth allows geophysicists to determine the density and stiffness of rocks at depths inaccessible to direct examination. If it is possible to match up those properties with the properties of known substances at elevated temperatures and pressures, it is possible (in principle) to infer what the environmental conditions must be deep in the earth.

The problem with this is that the conditions are so extreme at the earth's center that it is very difficult to perform any kind of laboratory experiment that faithfully simulates conditions in the earth's core. Nevertheless, geophysicists are constantly trying these experiments and improving on them, so that their results can be extrapolated to the earth's center, where the pressure is more than three million times atmospheric pressure.

The bottom line of these efforts is that there is a rather wide range of current estimates of the earth's core temperature. The "popular" estimates range from about 4,000 kelvins up to over 7,000 kelvins (about 7,000 to 12,000 degrees Fahrenheit). If we knew the melting temperature of iron very precisely at high pressure, we could pin down the temperature of the earth's core more precisely, because it is largely made up of molten iron. But until our experiments at high temperature and pressure become more precise, uncertainty in this fundamental property of our planet will persist.

?

What causes the regular, wavelike shapes that form in the sand on beaches?

ANSWERED BY:

Robert S. Anderson, Associate Professor of Earth Sciences, University of California at Santa Cruz.

Ripples in the sand, found on both beaches and dunes, are one of nature's most ubiquitous and spectacular examples of self-organization. They do not result from some predetermined pattern in the wind that is somehow impressed on the surface but rather from the dynamics of the individual grains in motion across the surface. They arise whenever wind blows strongly enough over a sand surface to entrain grains into the wind. The subsequent hopping and leaping of these grains is called saltation. Saltating grains travel elongated, asymmetric trajectories: Rising relatively steeply off the bed, their path is then stretched downwind as they are accelerated by drag forces. They impact the sand surface centimeters to tens of centimeters downwind, typically at a low angle, around 10 degrees. It is this beam of wind-accelerated grains impacting the sand surface at a low angle that is responsible for ripples.

An artificially flattened sand surface will not remain flat for long. (Try it on the beach or on the upwind side of a dune and see for yourself.) Small irregular mottles in the sand surface, perhaps a couple of centimeters in wavelength, rapidly arise and grow once the wind starts to blow hard enough to initiate saltation. They then slowly organize themselves into more regular waves whose low crests are aligned perpendicular to the wind direction and begin to march slowly downwind. Typical ripple spacing is about 10 centimeters, whereas the typical height of the crests above the troughs is a few millimeters. The pattern is never perfect, but instead the ripple crests occasionally split or terminate, generating a pattern that looks remarkably like one's fingerprint. In cross section, the ripples are asymmetric, having low-angle upwind faces and steeper downwind faces. Interestingly, the larger grains tend to accumulate on the crests of the ripples, leaving the troughs enriched in smaller grains.

What sets the pattern of deposition and erosion? Because of their low-impact angle, the intensity of the bombardment by energetic, long trajectories is highest on the upwind sides of the ripples. This results in many more grains being launched off from the upwind surface than are landing there, causing erosion. Conversely, the downwind face of the ripple is inclined so steeply that it does not get bombarded; the crest of the ripple in essence shields the downwind face from bombardment. The downwind face therefore becomes a zone of net accumulation; the many grains blasted from the upwind face and crest of the ripple land here, while few grains are blasted from the surface. This process explains why even small initial bumps in the sand surface will grow and why ripples move downwind.

And what controls the wavelength? Indirectly, the wind speed, in the following way. The wind speed determines the impact angle of the longer trajectories: The greater the wind speed, the more the saltation trajectories are stretched out, and the lower the resulting impact angles. A lower-impact angle results in a longer bombardment shadow behind each ripple crest and hence a longer wavelength.

?

What is quicksand?

ANSWERED BY:
**Darrel G. F. Long, Sedimentologist, Department of
Earth Sciences, Laurentian University, Sudbury,
Ontario, Canada.**

Quicksand is a mixture of sand and water or of sand and air; it looks solid but becomes unstable when it is disturbed by any additional stress. Grains frequently are elongated rather than spherical, so loose packing can produce a configuration in which the spaces between the granules, or voids, filled with air or water make up 30 to 70 percent of the total volume. This arrangement is similar to a house of cards, in which the space between the cards is significantly greater than the space occupied by the cards. In quicksand, the sand collapses, or becomes "quick," when force from loading, vibration, or the upward migration of water overcomes the friction holding the particles in place. In normal sand, in contrast, tight packing forms a rigid mass, with voids making up only about 25 to 30 percent of the volume.

Most quicksand occurs in settings where there are natural springs, such as at the base of alluvial fans (cone-shaped bodies of sand and gravel formed by rivers flowing from mountains), along riverbanks, or on beaches at low tide. Quicksand does appear in deserts, on the loosely packed, downwind sides of dunes, but this is rare. And the amount of sinking is limited to a few centimeters, because once the air in the voids is expelled, the grains nestle too close together to allow further compaction.

Let's Get Wet
oceans

?

How did the oceans form?

ANSWERED BY:
**Tobias C. Owen, Institute for Astronomy,
Honolulu, Hawaii.**

The origin of the oceans goes back to the time of the
earth's formation 4.6 billion years ago, when our
planet was forming through the accumulation of smaller
objects, called planetesimals. There are basically three possible sources for the water. It could have (1) separated out
from the rocks that make up the bulk of the earth; (2)
arrived as part of a late-accreting veneer of water-rich
meteorites; or (3) arrived as part of a late-accreting veneer
of icy planetesimals, that is, comets.

The composition of the ocean offers some clues as to
its origin. If all the comets contain the same kind of water
ice that we have examined in Comets Halley and Hyaku-
take—the only ones whose water molecules we've been
able to study in detail—then comets cannot have delivered

all the water in the earth's oceans. We know this because the ice in the comets contains twice as many atoms of deuterium (a heavy isotope of hydrogen) to each atom of ordinary hydrogen as we find in seawater.

At the same time, we know that the meteorites could not have delivered all of the water, because then the earth's atmosphere would contain nearly 10 times as much xenon (an inert gas) as it actually does. Meteorites all carry this excess xenon. Nobody has yet measured the concentration of xenon in comets, but recent laboratory experiments on the trapping of gases by ice forming at low temperatures suggest that comets do not contain high concentrations of xenon. A mixture of meteoritic water and cometary water would not work either, because this combination would still contain a higher concentration of deuterium than is found in the oceans.

Hence, the best model for the source of the oceans at the moment is a combination of water derived from comets and water that was caught up in the rocky body of the earth as it formed. This mixture satisfies the xenon problem. It also appears to solve the deuterium problem—but only if the rocky material out near the earth's present orbit picked up some local water from the solar nebula (the cloud of gas and dust surrounding the young sun) before they accreted to form the earth. Some new laboratory studies of the manner in which deuterium gets exchanged between gas and water vapor have indicated that the water vapor in the local region of the solar nebula would have had about the right (low) proportion of deuterium to the excess deuterium seen in comets.

?

Why does the ocean appear blue? Is it because it reflects the sky?

ANSWERED BY:
The Staff of Skidaway Institute of Oceanography, Savannah, Georgia.

The ocean looks blue because red, orange, and yellow light (long-wavelength light) are absorbed more strongly by water than is blue light (short-wavelength light). So when white light from the sun enters the ocean, it is mostly the blue that gets returned. The sky appears blue for the same reason.

In other words, the color of the ocean and the color of the sky are related but occur independently of each other: In both cases, the preferential absorption of long-wavelength (reddish) light gives rise to the blue light. Note that this effect only works if the water is very pure; if the water is full of mud, algae, or other impurities, the light scattered off these impurities will overwhelm the water's natural blueness.

6

Count on Me

Much Ado About Nothing
zero

?

What is the origin of zero?

ANSWERED BY:
Robert Kaplan, Author of *The Nothing That Is:
A Natural History of Zero.*

The first evidence we have of zero is from the Sumerian culture in Mesopotamia, some 5,000 years ago. There a slanted double wedge was inserted between the cuneiform symbols for numbers, written positionally, to indicate the absence of a number in a place (as we would write 102, the "0" indicating no digit in the tens column).

The symbol changed over time as positional notation, for which zero was crucial, made its way to the Babylonian empire and from there to India, via the Greeks (in whose own culture zero made a late and only occasional appearance; the Romans had no trace of it at all). Arab merchants brought the zero they found in India to the West, and after many adventures and much opposition, the symbol we use took hold and the concept flourished, as zero took on much more than a positional meaning and has played a crucial role in our mathematizing of the world.

The mathematical zero and the philosophical notion of nothingness are related but aren't the same. Nothingness plays a central role very early on in Indian thought (there called "sunya"), and we find speculation in virtually all cosmogonical myths about what must have preceded the world's creation. So in the Bible's book of Genesis (1:2): "And the earth was without form, and void."

But our inability to conceive of such a void is well caught in the book of Job, who cannot reply when God asks of him (Job 38:4): "Where wast thou when I laid the foundations of the earth? Declare, if thou hast understanding." Our own era's physical theories about the Big Bang cannot quite reach back to an ultimate beginning from nothing—although in mathematics we can generate all numbers from the empty set. Nothingness as the state out of which alone we can freely make our own natures lies at the heart of existentialism, which flourished in the mid-twentieth century.

Give 'Em an Inch
measurement

?

On average, how many degrees apart is any one person in the world from another?

ANSWERED BY:
Duncan J. Watts, Associate Professor of Sociology, Columbia University, New York City, and Author of *Six Degrees: The Science of a Connected Age*.

This is a question with rather a long history. As early as 1929, the Hungarian writer Frigyes Karinthy speculated that anyone in the world could be connected to anyone else through a chain consisting of no more than five intermediaries. Because the last person in the chain, who we call the target, does not count as an intermediary, five intermediaries are equivalent to six degrees of separation. The first scientific exploration of what was to become known as the "small-world problem" came almost three decades later in the work of Manfred Kochen (a mathematician) and Ithiel de Sola Pool (a political scientist), who proposed a mathematical explanation of the problem. Assuming that individuals choose 1,000 friends at random from a population as large as 100 million, Kochen and Pool showed that no more than two or three intermediaries (hence three or four degrees of separation) would be required to connect any two people. People, however, do not choose friends at random, which implies that the real answer should be higher. Kochen and Pool realized this but were unable to solve the more difficult problem.

Stimulated by Pool and Kochen's work, the great social psychologist Stanley Milgram devised an ingenious experiment in the late 1960s to test the hypothesis. Milgram and his graduate student Jeffrey Travers gave 300 letters to subjects in Boston and Omaha, with instructions to deliver them to a single target person (a stockbroker from Sharon, Massachusetts) by mailing the letter to an acquaintance who the subject deemed closer to the target. The acquaintance then got the same set of instructions, thus setting up a chain of intermediaries. Milgram found that the average length of the chains (64 of them) that completed the experiment was about six—quite remarkable in light of Karinthy's prediction 40 years earlier. Since Milgram, the small-world problem has become a cultural phenomenon, especially after the playwright John Guare chose the catchy term "six degrees of separation" as the title of his 1990 play. But until recently, very little empirical work had been done aside from Milgram's initial experiment, and no one could explain why it worked.

Some recent theoretical work suggests that the answer may or may not be 6, but it is certainly small—not 100, for example.

?

Where does the measurement of the "meter" come from?

ANSWERED BY:
Barry N. Taylor, Fundamental Constants Data Center, National Institute of Standards and Technology, Gaithersburg, Maryland.

The origins of the meter go back to the eighteenth century. At that time, there were two competing proposals for how to define a standard unit of measure, or meter.

The astronomer Christian Huygens suggested that the meter be defined by the length of a pendulum having a period of one second; others favored a meter defined as one ten-millionth the length of the earth's meridian along a quadrant (one-fourth the circumference of the earth). In 1791, soon after the French Revolution, the French Academy of Sciences endorsed the meridian definition because the force of gravity varies slightly over the surface of the earth, affecting the period of a pendulum.

Researchers measured the arc from Dunkirk, France, to Barcelona, Spain; on June 22, 1799, the French Academy Archives adopted its standard meter and recorded it on a platinum bar. (The French government made the meter the compulsory standard of measure in 1840.) The French, however, miscalculated the flattening of the earth due to its rotation. As a result, the meter in the Archives is 0.2 millimeter shorter than one ten-millionth of the quadrant of the earth.

Despite its flaws, the French definition of the meter stuck. The Treaty of the Meter was signed in 1875, and in 1889 a platinum-iridium bar was established as the International Prototype Meter. (It was selected from several candidate meters because it was the closest to the Meter of the Archives, the platinum bar held in the French Academy.)

The meter bar lasted a good long time; but it became cumbersome and error-prone to refer to a specific, physical meter bar. Finally, after 71 years, a new standard emerged.

In 1960 the General Conference on Weights and Measures redefined the meter in terms of the number of waves of a very precise color (wavelength) of light emitted by the element krypton [86 atoms]. That revision did not last very long. In 1983 the Conference discarded the krypton standard and redefined the meter in terms of the speed of light—what might be called a theoretical definition. The meter is now officially 1/299,792,458 the distance traveled by light in a vacuum in one second.

These changing definitions offer a good example of what happens in the field of measurement: As the tools and available precision change, the standards change with them. These days, you can buy a laser from Hewlett-Packard and create a reference meter on your own; the level of accuracy is far beyond what any scientist could achieve a century ago.

?

How does a laser measure the speed of a car?

ANSWERED BY:
James A. Worthey, Office of Law Enforcement Standards, National Institute of Standards and Technology, Gaithersburg, Maryland.

The laser speed gun used by the police contains a pulsed diode laser. When the police officer squeezes the trigger, the laser emits a brief pulse of infrared light, which is focused by a lens so that it travels as a narrow beam. The pulse reflects off the moving car; a small fraction of the original pulse energy is received by a second lens and focused onto a fast, sensitive detector. Electronic timing circuits measure the pulse's round-trip time of flight.

Multiplying the round-trip time by the speed of light in air and dividing by 2 (because of the round-trip) gives the distance, or "range," of the car. A few milliseconds later, the laser pulses for a new range measurement. The new range will be slightly less or more, depending on whether the car is moving toward the policeman or away. The laser speed gun continues to collect data in this way until dozens of range measurements have been made; the whole process takes only about half a second. The data are then analyzed by a computer in the speed gun. If the range changed steadily during the series of measurements, then the "slope of the graph"—that is, the change in distance between each pulse—indicates the motorist's speed. If the range data did not change steadily, this is an indication of error. If the computer is satisfied that the data are good, the speed gun displays the speed and range (distance) of the car.

The level of precision needed to catch a speeding automobile is quite remarkable. The speed of light in air is 299,705,663 meters per second. If the car is 150 meters (about 500 feet) away, then the round-trip travel time for the laser pulse is about one microsecond, that is, one-millionth of a second. To determine speed accurately, the speed gun must measure time with split-nanosecond (billionths of a second) accuracy.

Does Not Compute
computers

?

Why do computers crash?

ANSWERED BY:
Clay Shields, Assistant Professor of Computer Science, Georgetown University, Washington, D.C.

The short answer is: for many reasons. Computers crash because of errors in the operating system (OS) software or the machine's hardware. Software glitches are probably more common, but those in hardware can be devastating.

The OS does more than allow the user to operate the computer. It provides an interface between applications and the hardware and directs the sharing of system resources among different programs. Any of these tasks can go awry. Perhaps the most common problem occurs when, because of a programming flaw, the OS tries to access an incorrect memory address. In some versions of Microsoft Windows, users might see a general protection fault (GPF) error message; the solution is to restart the program or reboot the computer. Other programming mistakes can drive the OS into an infinite loop, in which it executes the same instructions over and over. The computer appears to lock up and must be reset. Another way things can go amiss: when a programming bug allows information to be written into a memory buffer that is too small to accept it. The information "overflows" out of the buffer and overwrites data in memory, corrupting the OS.

Application programs can also cause difficulties. Newer operating systems have built-in safeguards, but application bugs can affect older ones. Software drivers, which are added to the OS to run devices such as printers, may stir up trouble. That's why most modern operating systems have a special boot mode that lets users load drivers one at a time, so they can determine which is to blame.

Hardware components must also function correctly for a computer to work. As these components age, their performance degrades. Because the resulting defects are often transient, they are hard to diagnose. For example, a computer's power supply normally converts alternating current to direct current. If it starts to fail and generates a noisy signal, the computer can crash.

Errors on a computer's hard drive are the most intractable. Hard disks store information in units called sectors. If the sectors go bad, the data stored on them go, too. If these sectors hold system information, the computer can freeze up. Bad sectors also can result from an earlier crash. The system information becomes corrupted, making the computer unstable; ultimately the OS must be reinstalled. Last and worst, a computer can fail completely and permanently if the machine gets jarred and the head that reads information makes contact with the disk surface.

How do Internet search engines work?

ANSWERED BY:
Javed Mostafa, Victor H. Yngve Associate Professor of Information Research Science and Director of the Laboratory of Applied Informatics, Indiana University, Bloomington.

Publicly available Web services—such as Google, InfoSeek, Northern Light, and AltaVista—employ various techniques to speed up and refine their searches.

One way to save search time is to match the Web user's query against an index file of preprocessed data stored in one location, instead of sorting through millions of Web sites. To update the preprocessed data, software called a crawler is sent periodically by the database to collect Web pages. A different program parses the retrieved pages to extract search words. These words are stored, along with the links to the corresponding pages, in the index file. New user queries are then matched against this index file.

"Smart representation" refers to selecting an index structure that minimizes search time. Data are far more efficiently organized in a "tree" than in a sequential list. In an index tree, the search starts at the "top," or root node. For search terms that start with letters that are earlier in the alphabet than the node word, the search proceeds down a "left" branch; for later letters, "right." At each subsequent node there are further branches to try, until the search term is either found or established as not being on the tree. The URLs, or links, produced as a result of such searches are usually numerous. But because of ambiguities of language (consider "window blind" versus "blind ambition"), the resulting links would generally not be equally relevant. To glean the most pertinent records, the search algorithm applies ranking strategies. A common method, known as term-frequency-inverse document-frequency, determines relative weights for words to signify their importance in individual documents; the weights are based on the distribution of the words and the frequency with which they occur. Words that occur very often (such as "or, . . . to," and "with") and that appear in many documents have substantially less weight than do words that appear in relatively few documents and are semantically more relevant.

Link analysis is another weighting strategy. This technique considers the nature of each page—namely, if it is an "authority" (a number of other pages point to it) or a "hub" (it points to a number of other pages). The highly successful Google search engine uses this method to polish searches.

?

How do rewritable CDs work?

ANSWERED BY:
Gordon Rudd, President, Clover Systems, Laguna Hills, California.

All CDs, and DVDs, work by virtue of marks on the disc that appear darker than the background. These are detected by shining a laser on them and measuring the reflected light. In the case of molded CDs or DVDs, such as those bought in music or video stores, these marks are physical "pits" imprinted into the surface of the disc. In CD-Recordable (CD-R) discs, a computer's writing laser creates permanent marks in a layer of dye polymer in the disc.

CD-Rewritable (CD-RW) discs are produced in a similar fashion, except that the change to the recording surface is reversible. The key is a layer of phase-change material, an alloy composed of silver, indium, antimony, and tellurium. Unlike most solids, this alloy can exist in either of two solid states: crystalline (with atoms closely packed in a rigid and organized array) or amorphous (with atoms in random positions). The amorphous state reflects less light than the crystalline one does.

When heated with a laser to about 700 degrees Celsius, the alloy switches from the original crystalline phase to the amorphous state, which then appears as a dark spot when the disc is played back. These spots can be erased using the same laser (at a lower power) to heat the material to a temperature of about 200 degrees Celsius; this process returns the alloy to its crystalline state. Most CD-RW makers suggest that one disc can be overwritten up to 1,000 times and will last about 30 years.

?

When did the term "computer virus" arise?

ANSWERED BY:

Rob Rosenberger, a computer consultant who maintains the Computer Virus Myths homepage.

The roots of the modern computer virus go back to 1949, when the computer pioneer John von Neumann presented a paper on the "Theory and Organization of Complicated Automata," in which he postulated that a computer program could reproduce. Bell Labs employees gave life to von Neumann's theory in the 1950s in a game they called "Core Wars." In this game, two programmers would unleash software "organisms" and watch as they vied for control of the computer.

Strangely enough, two science-fiction books in the 1970s helped to promote the concept of a replicating program. John Brunner's *Shockwave Rider* and Thomas Ryan's *Adolescence of P-1* depicted worlds where a piece of software could transfer itself from one computer to another without detection. Back in the real world, Fred Cohen presented the first rigorous mathematical definition for a computer virus in his 1986 Ph.D. thesis. Cohen coined the term "virus" at this point and is considered the father of what we know today as a computer virus. He sums it up in one sentence as "a program that can infect other programs by modifying them to include a, possibly evolved, version of itself."

The media seldom mentioned computer viruses in the mid-1980s, treating the whole concept as an obscure theoretical problem. The media's perception of viruses took a dramatic turn in late 1988, when a college student named Robert T. Morris unleashed the infamous "Internet Worm." (Some trivia: Morris's father had a hand in the original Core Wars games.) Reporters grew infatuated with the idea of a tiny piece of software knocking out big mainframe computers worldwide. The rest, as they say, is history.

Let's Get Physical

PHYSICS

Let There Be Light
light

?

How do surfaces, such as pavement, become
heated from the sun?

ANSWERED BY:
Scott M. Auerbach, Theoretical Chemist, University of
Massachusetts, Amherst.

The simple answer is that light from the sun excites electrons in the atoms that constitute surfaces such as pavement. Here's how it works: The atoms of the pavement are perpetually vibrating. Some of those atoms vibrate sufficiently vigorously that their vibrational energy is roughly equal to the electronic energy absorbed from the sun—in essence, they are in resonance with the solar energy. Those atoms then make a quantum transition from "electronically excited" to "vibrationally excited," meaning that the energy causes the whole atom to move. We feel that motion as "heat." The atoms that make the jump to vibrational excitation soon collide into neighboring atoms, dissipating their vibrational energy throughout the surface, making the surface hot throughout.

What is the physical process by which a mirror reflects light rays?

ANSWERED BY:
Peter N. Saeta, Assistant Professor of Physics,
Harvey Mudd College, Claremont, California.

B ecause there are a great many electrons in the mirror, all vibrating at the frequency of the incident light, reflection from the mirror is really a group effort. All the electrons dance to the same music, whose rhythm is provided by the incident light wave. This coordination causes the reflected wave to make the same angle with respect to the mirror's surface as does the incident beam.

A typical mirror consists of a piece of glass that has been coated with a layer of metal. Glass by itself reflects a little of the light, but the metal layer greatly boosts the reflectivity. If the metal were perfectly conducting, it would reflect all of the light, but the conductivity of real metals is less than perfect. This imperfection leads to some absorption of light in the metal. A polished silver surface, for example, reflects about 93 percent of the incident visible light, which is very good as metals go. Interestingly, if the metal layer is very thin—only a few hundred atoms thick—then much of the light leaks through the metal and comes out the back. If you get the thickness of a metal layer right, you can make a beam splitter that divides an incident beam of light into two equal parts, with just a little bit of the light lost to the metal film itself.

As good as the reflectivity of a silver mirror is, you can do much better with dielectric mirrors. These reflectors consist of alternating layers of two transparent materials that have different indices of refraction. Dielectric mirrors can have reflectivities of 99.999 percent or better at the wavelength for which they are designed. In these mirrors, essentially all of the incident light reflects, and virtually none is absorbed in the mirror or transmitted through it.

?

How does sunscreen protect the skin?

ANSWERED BY:
John Sottery, a leading sunscreen researcher, and President, IMS, Inc., Milford, Connecticut.

N atural sunlight contains, among other things, ultra-violet (UV) photons. These photons are shorter in wavelength and higher in energy than visible light. Because they fall outside the visible spectrum, the human eye cannot perceive them. When it comes to sun exposure, however, what you can't see will hurt you. When these high-energy photons strike your skin, they generate free radicals and can also directly damage your DNA. Over the short term, this UV-induced damage can produce a painful burn; over the long term, it causes premature aging of the skin, as well as millions of new cases of skin cancer each year.

The UV rays that we are exposed to here on the earth's

surface consist of UVB and UVA photons. The shorter wavelength UVB rays don't penetrate deeply into the skin; they cause significant damage to DNA and are the primary cause of sunburn and skin cancer. The longer wavelength UVA rays penetrate the deeper layers of the skin, where they produce free radicals. UVA exposure has been linked to premature aging of the skin and immunologic problems.

A sunscreen product acts like a very thin bulletproof vest, stopping the UV photons before they can reach the skin and inflict damage. It contains organic sunscreen molecules that absorb UV and inorganic pigments that absorb, scatter, and reflect UV. To deliver a high level of protection, a sunscreen product must have sufficient quantities of these protective agents and it must optimally deploy them over the skin's peaks and valleys.

The term SPF that appears on sunscreen labels stands for sun protection factor, but it is really a sunburn protection factor. Products with a higher SPF allow fewer of the photons that produce sunburn to strike the skin. In simple terms, you can view an SPF 10 sunscreen as allowing 10 out of every 100 photons to reach the skin and an SPF 20 product as allowing only 5 out of every 100 photons to reach the skin. Because sunburn is primarily a UVB effect, it is possible for a sunscreen product to deliver high SPF while allowing a significant percentage of the incident UVA photons to reach the skin. To deliver true broad spectrum protection, products must also block a significant fraction of the UVA photons. In the U.S. market, this requires that the products contain significant levels of zinc oxide, avobenzone, or titanium dioxide.

In the case of tanning beds, the UV output differs from bed to bed, but it generally contains less UVB and significantly more UVA than does natural sunlight. This leads to less sunburn and more tanning. In the long term, however, the UVA rays take their toll on skin, and tanning beds do not represent a safe tanning option.

?

Why are sunsets orange?

ANSWERED BY:
Michael Kruger, Department of Physics, University of Missouri, Kansas City.

When the sun is setting, the light that reaches you has had to go through lots more atmosphere than when the sun is overhead; hence the only color light that is not scattered away is the long wavelength light, the red. We can also answer why clouds, milk, powdered sugar, and salt are white. The particles in these materials that are responsible for scattering the light are larger than the wavelength of light. Consequently, all colors of light are scattered by more or less the same amount. Much of the scattering in milk is due to the lipids (fat). If you take out the fat, the milk will not scatter as much light; that is probably why skim milk looks the way it does.

I'm Very Particular
particles

?

If we cannot see electrons and protons, or smaller particles such as quarks, how can we be sure they exist?

ANSWERED BY:
Stephen Reucroft and John Swain, Professors of Physics, Northeastern University, Boston, Massachusetts.

The central concept here is what we mean by "see." Normally when we say that we "see" an object, what we mean is that we detect with our eyes particles of light called photons, which come from some source like a lightbulb or the sun and bounce off the object. The idea of being able to "see" things by observing particles that scatter from them is common to particle physics experiments that study tiny objects like electrons, protons, and quarks (out of which protons are made).

For a classical physics picture of how this works, you might imagine being in a large dark room with an object whose shape you don't know. If you have a bucket of tennis balls, you can start to build up a picture of what the object looks like by tossing the balls at it. Instead of tennis balls, particle physicists use small particles, such as electrons at very high energies. Recalling Albert Einstein's famous formula, $E = mc^2$, lots of energy can be traded for a little bit of matter.

So when we use very high energies to see deeply into matter, a new phenomenon can occur: Not only do we see what's there, but we create matter in the form of new particles. This gives us yet another way to study the structure of the world at its most profound level.

?

Is glass really a liquid?

ANSWERED BY:

Steve W. Martin, Associate Professor of Materials Science and Engineering, Iowa State University, Ames.

The seeming paradox that a glass is at the same time a liquid and a solid is not easily reconciled. Glasses are "solids" produced by cooling a molten liquid fast enough that crystallization does not occur at the normal freezing point. Instead, the liquid supercools into the thermodynamic never-never land of metastability: kinetically settled enough to exist as a well-defined state of matter, yet not truly thermodynamically stable. As the supercooled liquid cools to lower and lower temperatures, the viscosity of the liquid increases dramatically. That happens because as thermal energy becomes ever less available, chemical bonds within the liquid constrain the atomic motion more and more.

As the glass cools, the time it needs to demonstrate liquid behavior (the "viscous relaxation time") increases and eventually reaches extremes. On a short timescale, the "liquid" glass will appear solid, but after a short while, it can be seen to be slowly flowing, like incredibly thick syrup. At still lower temperatures, the relaxation time reaches values that are truly geologic, i.e., many millions of years. Window glass at room temperature has a nearly incalculable relaxation time, approaching the age of the universe itself. For all practical observations, this glass is a solid. But its solidity is in the eye of the beholder.

Now Hear This
sound

?

How can the extremity of a whip travel faster than the speed of sound to produce the characteristic "crack"?

ANSWERED BY:

Collier Smith, National Institute of Standards and Technology, Gaithersburg, Maryland.

Looking at how the whip behaves gives a clue to this phenomenon (and it helps if you have actually tried to crack a whip). The whip has to be moved so that a U-shaped loop is formed near the handle, where the whip is thickest and stiffest. As the whip is swung, the loop travels outward toward the thinner, lighter tip. The loop travels progressively faster the closer it gets to the tip, because the energy from the heavier part of the whip is carried along into the lighter, thinner part. This amplification is analogous to the way in which an ocean wave of small height becomes a high breaker as it enters the shallow water near the shore or over a reef.

When the loop reaches the end, it is going extremely fast and causes the very tip to "whip" around in a tight circle. It is at this point that the tip exceeds the speed necessary to create a tiny shock wave in the air; it is this shock that we hear as the "crack."

?

What causes the noise emitted from high-voltage power lines?

ANSWERED BY:
Robert Dent, President, IEEE Power Engineering Society, Piscataway, New Jersey.

The noise that we hear emitted from high-voltage lines is caused by the discharge of energy that occurs when the electrical field strength on the conductor surface is greater than the "breakdown strength" (the field intensity necessary to start a flow of electric current) of the air surrounding the conductor. This discharge is also responsible for radio noise, a visible glow of light near the conductor, an energy loss known as corona loss, and other phenomena associated with high-voltage lines.

The degree or intensity of the so-called corona discharge and the resulting noise are affected by the condition of the air, that is, by humidity, air density, wind, and water in the form of rain, drizzle, and fog. Water increases the conductivity of the air and so increases the intensity of the discharge. Also, irregularities on the conductor surface, such as nicks or sharp points and airborne contaminants, can increase the corona activity. Aging or weathering of the conductor surface generally reduces the significance of these factors.

The higher voltages at which modern lines operate have increased the noise problem to the point to which they have become a concern to the power industry. Consequently, these lines are now designed, constructed, and maintained so that during dry conditions they will operate below the corona-inception voltage, meaning that the line will generate a minimum of corona-related noise. In bad weather conditions, however, corona discharges can be produced by water droplets, fog, and snow.

?

What are "booming sands" and what causes
the sounds they make?

ANSWERED BY:
Thomas S. Ahlbrandt, U.S. Geological Survey,
Denver, Colorado.

For anyone who has experienced these phenomena, as I
have in several North American deserts, the mysticism
or romance of the desert is amplified. Booming dunes have
been discussed in Middle East literature for at least 1,500
years; in Chinese literature for 1,200 years. The British
investigator R. A. Bagnold, who wrote a pioneering book in
1941, updated in 1954, entitled *The Physics of Blown Sand
and Desert Dunes,* described them as the "song of the sands."

Dunes are known to make two very different music-
like sounds when sand is sheared (commonly by the avalanche process). These sounds fall broadly into either the
"whistling sand" or "booming sand" category. A higher
pitched sound in the 800 to 1,200 Hz or 500 to 2,500 Hz
range is known to occur in dune sand and beach sand; most
commonly in dry sand but occasionally in wet sand conditions. Sands making such sounds are known as whistling,
singing, squeaking, or barking sands. The sound usually

lasts for a very short time, less than a second. "Booming sands" are much more impressive. One both hears them in a lower frequency band (50 to 264 Hz) than "whistling sand" and feels them—the ground trembles; the surface moves and ripples.

To understand how sand avalanches, I need to briefly explain how sand is moved by wind. When wind reaches a certain threshold speed (generally at about 14 mph) as it flows over loose sand, the sand starts to move in a series of jerks and jumps called saltation. As wind speed increases, saltation height increases to the point that one can truly have a miserable experience even at eye level. The grains bounce high due to the impact of one grain against another, like billiard balls, but sand grains also are lifted aerodynamically by the spin imparted to the grain by the striking grain.

The moving sand starts accumulating into a sand dune, and the dune builds to the point where the downwind side of the dune actually is sheltered from the wind. Sand is dropped from saltation (a process known as grainfall) and accumulates until it exceeds the angle of repose of sand, generally from 32 to 34 degrees. At this point, dry sand will start to avalanche. My personal experience is that sands boom after a strong sandstorm where avalanching, a continuous process during such storms, has not stabilized. Thus, there is an unstable condition where an equilibrium has yet to develop; any change, such as the sudden abatement of wind, may result in "booming" if the sand has the right characteristics.

Some researchers have carefully examined the textures and surfaces of booming sands and concluded that the sand grains need to be highly polished and moderately well rounded. This high degree of polishing, sphericity, and sorting is produced in certain dune settings as grains saltate into one another, each impact further rounding and polishing the grains. Many sands do not "boom" due to surface texture, grain size ranges, or the occurrence of moisture and vegetation.

As a final thought, some authors conclude that although booming sand is relatively uncommon on Earth, it may be common in the waterless or near waterless environments of the moon and Mars. One wonders if the *Sojourner* rover has felt or heard the "song of the sands."

What happens when an aircraft breaks the sound barrier?

ANSWERED BY:
Tobias Rossmann, Research Engineer, Advanced Projects Research, Inc., La Verne, California, and Visiting Researcher, California Institute of Technology, Pasadena.

A discussion of what happens when an object breaks the sound barrier must begin with the physical description of sound as a wave with a finite speed. Anyone who has been far enough away from an event to see it first and then hear it is familiar with the relatively slow speed of sound waves. At sea level and a temperature of 22 degrees Celsius, sound waves travel at 345 meters per second (770 mph). As the temperature decreases, the sound speed also drops, so that for a plane flying at 35,000 feet—where the temperature is, say, -54 degrees Celsius— the local speed of sound is 295 meters per second (660 mph).

Because the speed of sound waves is finite, the sources of sound that are moving can begin to catch up with the sound waves they emit. As the speed of the object increases to the sonic velocity, sound waves begin to pile up in front of the object. If the object has sufficient acceleration, it can burst through this barrier of unsteady sound waves and jump ahead of the radiated sound, thus breaking the sound barrier.

An object traveling at supersonic speeds generates steady pressure waves that are attached to the front of the object (a bow shock). An observer hears no sound as an object approaches. After the object has passed, these generated waves (Mach waves) radiate toward the ground, and the pressure difference across them causes an audible effect, known as a sonic boom.

In Theory
theoretical physics

?

Is it theoretically possible to travel through time?

ANSWERED BY:
William A. Hiscock, Professor of Physics, Montana State University, Bozeman.

To answer this question, we must be a bit more specific about what we mean by traveling through time. Discounting the everyday progression of time, the question can be divided into two parts: Is it possible, within a short time (less than a human life span), to travel into the distant future? And is it possible to travel into the past? Our current understanding of fundamental physics tells us that the answer to the first question is a definite yes, and to the second, maybe.

The mechanism for traveling into the distant future is to use the time-dilation effect of special relativity, which states that a moving clock appears to tick more slowly the closer it approaches the speed of light. This effect, which has been overwhelmingly supported by experimental tests, applies to all types of clocks, including biological aging.

If one were to depart from the earth in a spaceship that could accelerate continuously at a comfortable one g (an acceleration that would produce a force equal to the gravity at the earth's surface), one would begin to approach the speed of light relative to the earth within about a year. As the ship continued to accelerate, it would come ever closer to the speed of light, and its clocks would appear to run at an ever slower rate relative to the earth. Under such circumstances, a round trip to the center of our galaxy and back to the earth—a distance of some 60,000 light-years—could be completed in only a little more than 40 years of ship time. Upon arriving back at the earth, the astronaut would be only 40 years older, while 60,000 years would have passed on the earth.

Such a trip would pose formidable engineering problems: The amount of energy required, even assuming a perfect conversion of mass into energy, is greater than the planetary mass. But nothing in the known laws of physics would prevent such a trip from occurring.

Time travel into the past, which is what people usually mean by time travel, is a much more uncertain proposition. There are many solutions to Einstein's equations of general relativity that allow a person to follow a timeline that would result in her (or him) encountering herself—or her grandmother—at an earlier time. The problem is deciding whether these solutions represent situations that could occur in the real universe or whether they are mere mathematical oddities incompatible with known physics. No experiment or observation has ever indicated that time travel is occurring in our universe. Much work has been done by theoretical physicists in the past decade to try to determine whether, in a universe that is initially without time travel, one can build a time machine—in other words, if it is possible to manipulate matter and the geometry of space-time in such a way as to create new paths that circle back in time.

How could one build a time machine? The simplest way currently being discussed is to take a wormhole (a tunnel connecting spatially separated regions of space-time) and give one mouth of the wormhole a substantial velocity with respect to the other. Passage through the wormhole would then allow travel to the past.

Easily said—but where does one obtain a wormhole? Although the theoretical properties of wormholes have been extensively studied over the past decade, little is known about how to form a macroscopic wormhole, large enough for a human or a spaceship to pass through. Some speculative theories of quantum gravity tell us that space-time has a complicated, foamlike structure of wormholes on the smallest scales—of 10^{33} centimeter, or a billion times smaller than an electron. Some physicists believe it may be possible to grab one of these truly microscopic wormholes and enlarge it to a usable size, but at present these ideas are all very hypothetical.

Even if we had a wormhole, would nature allow us to convert it into a time machine? Stephen Hawking has formulated a "Chronology Protection Conjecture," which states that the laws of nature prevent the creation of a time machine. At the moment, however, this is just a conjecture, not proven.

Theoretical physicists have studied various aspects of physics to determine whether this law or that might protect chronology and forbid the building of a time machine. In all the searching, however, only one bit of physics has been found that might prohibit using a wormhole to travel through time. In 1982, Deborah A. Konkowski of the U.S. Naval Academy and I showed that the energy in the vacuum state of a massless quantized field (such as the photon) would grow without bound as a time machine is being turned on, effectively preventing it from being used. Later studies have shown that it is unclear whether the growing energy would change the geometry of space-time rapidly enough to stop the operation of the time machine. Recent work has shown that the energy in the vacuum state of a field having mass (such as the electron) does not grow to unbounded levels; this finding indicates that there may be a way to engineer the particle physics to allow a time machine to work.

Perhaps the biggest surprise of the work of the past decade is that it is not obvious that the laws of physics forbid time travel. It is increasingly clear that the question may not be settled until scientists develop an adequate theory of quantum gravity.

?

Is dark matter theory or fact?

ANSWERED BY:
**Rhett Herman, Radford University, Radford, Virginia,
and Shane L. Larson, Montana State University,
Bozeman.**

Dark matter is just what its name implies; it is matter (or mass) in the universe that we cannot see directly using any of our telescopes. Our telescopes see not only visible radiation (constituting the spectrum of colors that our own eyes can detect) but other types of radiation as well.

Dark matter emits no infrared radiation, nor does it give off radio waves, ultraviolet radiation, x-rays, or gamma rays. It is truly "dark." Cosmologists believe we can only see about 10 percent of the matter in the universe. Until they can accurately determine the mass of the universe, they will not know for sure whether it is expanding infinitely or will stop expanding at some point and collapse.

How, then, can we say with confidence that we know dark matter exists? The way in which dark matter reveals its presence to us is through the gravitational effect it exerts on luminous matter in the universe. ("Luminous" matter is the matter we can see with our telescopes.) The most obvious example of the gravitational effects of dark matter can be observed when looking at the rotation of galaxies.

To study galactic rotation, astronomers look at the emission line spectra of stars in each part of the galaxy. When the light from a star is observed using a diffraction grating or a prism, the starlight is separated into its true colors, in much the same way ordinary sunlight can be separated into the full rainbow of colors known as the visible spectrum.

The true colors constituting starlight separate into a series of light and dark lines in the visible spectrum, with each colored line corresponding to a specific wavelength of light. The specific wavelengths at which these lines occur are characteristic of the elements the stars contain. Thus, they can be used as an elemental "fingerprint" to identify a star's composition.

When a star emitting these line spectra is moving away from us, all of the wavelengths of the spectral lines are shifted to higher values than they would have been were the star stationary or moving side to side (neither toward nor away from us). All of the spectral lines are thus shifted toward the long wavelength part of the spectrum or to the red end of the spectrum.

This shifting of the lines, known as a Doppler shift, toward the red end of the visible spectrum is the origin of the term "redshift." When a star has part of its motion toward us, the spectral lines are shifted to shorter wavelengths, or "blueshifted," toward the blue end of the spectrum. By measuring the shift in wavelength, researchers can calculate the precise speed of a star, either toward us or away from us.

When a galaxy is rotating, the starlight from stars on the side of the galaxy that is moving toward us is blueshifted, while the starlight from the stars on the other side of the galaxy is redshifted. Thus, we can tell how fast and in what direction each individual star in the galaxy is orbiting about the center of the galaxy.

When stars orbit the center of a galaxy, their orbital speed is determined by the distribution of the mass contained within the galaxy. A graph showing the orbital speeds of the stars versus their distances from the center of the galaxy is known as the "rotation curve" for the stars in the galaxy.

If one takes all the luminous matter that can be seen in the galaxy (stars, gas, and dust) and predicts the rotation curve using the well-known laws of gravitational physics discovered by Newton, the speed of stars should decrease in a predictable manner the farther away they are from the center of the galaxy.

Looking at the rotation curves of galaxies, however, astronomers have found that rotational speeds do not fall off with distance as expected. Instead, the curves level off, and stars far away from the center of the galaxy move faster than expected. The only way to account for this observation is that a large quantity of matter that cannot be seen—dark matter—exists in the galaxies.

?

Would you fall all the way through a theoretical hole in the earth?

ANSWERED BY:

Mark Shegelski, Associate Professor of Physics, University of Northern British Columbia, Prince George, Canada.

The simple answer is, theoretically, yes. First, let us ignore friction, the rotation of the earth, and other complications, and focus on the case of a hole or tunnel entering the earth at one point, going straight through its center, and coming back to the surface at the opposite side of the planet. If we treat the mass distribution in the earth as uniform, one would fall into the tunnel and then come back up to the surface on the other side in a manner much like the motion of a pendulum swinging down and up again. Assuming that the journey began with zero initial speed (simply dropping into the hole), your speed would increase and reach a maximum at the center of the earth, and then decrease until you reached the surface, at which point the speed would again be zero. The gravitational force exerted on the traveler would be proportional to his distance from the center of the earth: It's at a maximum at the surface and zero at the center. The total time required for this trip would be about 42 minutes. If there were no friction, there would be no energy loss so our traveler could oscillate into and out of the tunnel.

This trip could not take place in the real world for a number of reasons, including the implausibility of building a tunnel 12,756 kilometers long, displacing all of the material in the tunnel's proposed path, and having the tunnel go through both the earth's molten outer core and its inner core, where the temperature is about 6,000 degrees! It would be much easier to build such a tunnel in a small asteroid. Interestingly enough, for a tunnel that reaches from one point to another point on the earth's surface but does not pass through the center of the planet, the travel time would still be about 42 minutes. The reason for this is that although the tunnel is shorter, the gravitational force along its path is also decreased as compared to that of a tunnel that goes through the center of the planet, which means you would travel more slowly. Because the distance and the component of gravity decrease by the same factor, the travel time ends up being the same.

?

What is antimatter?

ANSWERED BY:
R. Michael Barnett, Lawrence Berkeley National Laboratory, Berkeley, California; and Helen R. Quinn, Stanford Linear Accelerator Center, Menlo Park, California.

In 1930 Paul Dirac formulated a quantum theory for the motion of electrons in electric and magnetic fields, the first theory that correctly included Einstein's theory of special relativity in this context. This theory led to a surprising prediction: The equations that described the electron also described, and in fact required, the existence of another type of particle with exactly the same mass as the electron but with a positive instead of a negative electric charge. This particle, which is called a positron, is the antiparticle of the electron, and it was the first example of antimatter.

Its discovery in experiments soon confirmed the remarkable prediction of antimatter in Dirac's theory. A cloud chamber picture taken by Carl D. Anderson in 1931 showed a particle entering a lead plate from below and passing through it. The direction of the curvature of the path, caused by a magnetic field, indicated that the particle was a positively charged one but with the same mass and other characteristics as an electron.

Dirac's prediction applies not only to the dectron but to all the fundamental constituents of matter (particles). Each type of particle must have a corresponding antiparticle type. The mass of any antiparticle is identical to that of the particle. All the rest of its properties are also closely related but with the signs of all charges reversed. For example, a proton has a positive electric charge, but an antiproton has a negative electric charge.

There is no intrinsic difference between particles and antiparticles; they appear on essentially the same footing in all particle theories. But there certainly is a dramatic difference in the numbers of these objects we find in the world around us. All the world is made of matter, but any anti-matter we produce in the laboratory soon disappears because it meets up with and is annihilated by matter particles.

Modern theories of particle physics and of the evolution of the universe suggest, or even require, that antimatter and matter were once equally common during the universe's earliest stages. Scientists are now attempting to explain why antimatter is so uncommon today.

?

Does the speed of light ever change?

ANSWERED BY:
Robert Ehrlich, Department of Physics, George Mason University, Fairfax, Virginia.

Scientists do continually try to measure the constants of physics, but the usual motivation is to get more precise values, rather than to check whether those values have changed. If you had a technique to measure the speed of light that was no better than one that was previously used, you might not bother to make another measurement unless there were some reason to believe the previous value was in error. Moreover, if you did measure the speed of light and got an anomalous value, the odds are good that you would conclude you had made a mistake in the measurement, rather than that the speed of light "hiccuped" that particular day. State-of-the-art measurement techniques are extremely complex and require all sorts of checks to be sure they have been performed correctly. If in fact the constants of nature were changing, it would be very difficult to know how precise your experiment needed to be to detect such a change, unless you had some way of estimating the expected rate of change. Random, occasional hiccups in the constants almost certainly would go undetected.

The physicist Paul A. M. Dirac once suggested that the universal gravitational constant, G, which measures the strength of attraction every particle of matter feels for every other particle, was actually weakening with time, in proportion to the age of the universe since the Big Bang. According to Dirac's theory, the force of gravity would be only half as strong in 10 billion years as it is today. Given this specific prediction of the rate of weakening, scientists could make a specific test. The moon's distance from the earth is understood to be slowly increasing because of its tidal interaction with the earth (as tidal friction slows the rotation of the earth, the size of the moon's orbit must increase to conserve the total angular movement of the earth-moon system). If gravity were growing weaker with time, the moon would recede from the earth even faster than conventional theory predicted. Precise timing measurement of laser pulses from the earth bounced off reflectors that the *Apollo* astronauts left on the moon shows that the moon is indeed slowly receding but only at the expected rate, not the faster one Dirac's theory would imply.

You Won't Believe Your Eyes
the physics of seeing

?

Why do beautiful bands of color appear in the tiny oil slicks that form on puddles?

ANSWERED BY:

Dinesh O. Shah, Charles A. Stokes Professor of Chemical Engineering and Anesthesiology, University of Florida, Gainesville, Florida.

Small amounts of oil are usually present on the road surface (for instance, lubricating oil from cars, trucks, and bicycles). When it rains, drops of oil float on the layer of water that collects on the road because the density of oil is less than that of water—the same reason that wood floats on water. Commercial oil formulations usually contain a surfactant, an additive that causes the oil drops to spread out into a thin film atop the water. That film is thickest in the center of the patch, or oil slick, and thinnest at the periphery.

Light reflects upward both from the top of the oil film and from the underlying interface between the oil and the water; the path length (the distance from the reflection to your eye) is slightly different depending on whether the returned light comes from the top or from the bottom of the oil film. If the difference in path length is an integral multiple of the wavelength of the light, rays reflected from the two locations will reinforce each other, a process called constructive interference. If, however, the rays reach your eye out of step, they will cancel each other out due to destructive interference.

Sunlight contains all the colors of the rainbow. Each color of light has a different wavelength. Hence, a given disparity in the path length will cause constructive interference of certain colors, whereas other colors will not be observed because of destructive interference. Because the oil film gradually thins from its center to its periphery, different bands of the oil slick produce different colors.

Why is it that when you look at the spinning propeller of a plane or fan, at a certain speed, the blades seem to move backward?

ANSWERED BY:
Collier N. Smith, Technical Writer, Boulder Laboratories, National Institute of Standards and Technology, Gaithersburg, Maryland.

This phenomenon is not common in direct vision, but it is often seen in movies and television. Film and TV actually consist of series of still photographs shown in rapid succession to fool the eye into seeing motion. The individual pictures do not actually move. Instead, the position of a moving object shifts with respect to the background in each successive still frame; when viewed rapidly, one gets an impression of smooth motion.

The backward motion illusion occurs when the speed of rotation is such that, in the interval between each frame, a new blade moves nearly into the position occupied by a blade in the previous frame. If the timing is precise, it looks like the propeller did not turn at all.

But if the blade interval doesn't exactly match the camera interval (or a multiple of it), then the propeller will seem to turn slowly forward or backward. When the next blade is a little slow in reaching the previous blade's position, the rotation appears to be backward; when the next blade arrives a fraction early, the visual impression is that the blades are turning forward.

When the blade interval and frame interval are quite different from each other, the blades become blurred and the phenomenon disappears.

?

Why do jets leave a white trail in the sky?

ANSWERED BY:

**Jenn Stroud Rossmann, Assistant Professor of
Engineering, Harvey Mudd College,
Claremont, California.**

Jets leave white trails, or contrails, in their wakes for the same reason you can sometimes see your breath. The hot, humid exhaust from jet engines mixes with the atmosphere, which at high altitude is of much lower vapor pressure and temperature than the exhaust gas. The water vapor contained in the jet exhaust condenses and may freeze, and this mixing process forms a cloud very similar to the one your hot breath makes on a cold day.

Jet engine exhaust contains carbon dioxide, oxides of sulfur and nitrogen, unburned fuel, soot, and metal particles, as well as water vapor. The soot provides condensation sites for the water vapor. Any particles present in the air provide additional sites.

Depending on a plane's altitude, and the temperature and humidity of the atmosphere, contrails may vary in their thickness, extent, and duration. The nature and persistence of jet contrails can be used to predict the weather.

A thin, short-lived contrail indicates low-humidity air at high altitude, a sign of fair weather; whereas a thick, long-lasting contrail reflects humid air at high altitudes and can be an early indicator of a storm.

The mixing gases contained in the contrail rotate with respect to the ambient air. These regions of rotating flow are called vortices. (Any sharp surface, such as the tip of a wing, can cause vortical flow in its wake if it is sufficiently large or the flow is sufficiently fast.) On occasion, these trailing vortices may interact with one another.

Recent research has suggested that the ice clouds contained in contrails cause greenhouse effects and contribute to global warming as part of the insulating blanket of moisture and gases in the atmosphere. Researchers in this area seized on the opportunity presented on September 11 and 12, 2001, over the United States. The complete cessation of commercial air traffic offered a control sky without contrails for use in quantifying the environmental effects of contrails.

Shake It Up
everyday physics

?

Does hot water freeze faster than cold water?

ANSWERED BY:

Takasama Takahashi, Physicist, St. Norbert College, De Pere, Wisconsin.

Cold water does not boil faster than hot water. The rate of heating of a liquid depends on the magnitude of the temperature difference between the liquid and its surroundings (the flame on the stove, for instance). As a result, cold water will absorb heat faster while it is still cold; once it warms up to the temperature of hot water, the heating rate slows down, and from there it takes just as long to bring it to a boil as the water that was hot to begin with. Because it takes cold water some time to reach the temperature of hot water, cold water clearly takes longer to boil than hot water does. There may be some psychological effect at play; cold water starts boiling sooner than one might expect because of the aforementioned greater heat absorption rate when water is colder.

To the first part of the question—"Does hot water freeze faster than cold water?"—the answer is "Not usually but possibly under certain conditions." It takes 540 calories to vaporize one gram of water, whereas it takes 100 calories to bring one gram of liquid water from 0 degrees Celsius to 100 degrees Celsius. When water is hotter than 80 degrees Celsius, the rate of cooling by rapid vaporization is very high because each evaporating gram draws at least 540 calories from the water left behind. This is a very large amount of heat compared with the one calorie per Celsius degree that is drawn from each gram of water that cools by regular thermal conduction.

It all depends on how fast the cooling occurs, and it turns out that hot water will not freeze before cold water but will freeze before lukewarm water. Water at 100 degrees Celsius, for example, will freeze before water warmer than 60 degrees Celsius but not before water cooler than 60 degrees Celsius. This phenomenon is particularly evident when the surface area that cools by rapid evaporation is large compared with the amount of water involved, such as when you wash a car with hot water on a cold winter day.

?

How does a microwave oven cook foods?

ANSWERED BY:

**Chad Mueller, Assistant Professor of Chemistry,
Birmingham-Southern College, Alabama.**

A microwave oven cooks food because the water molecules inside it absorb the microwave radiation and thereby heat up and heat the surrounding food. Microwave radiation will similarly heat up skin and other body parts. In fact, people stationed at big microwave towers in cold climates used to stand in front of the microwave generators in order to warm themselves. The radiation is harmful mostly to the parts of the body that cannot conduct the heat away very effectively—the eyes especially. That heat transfer could explain why one sometimes hears about people (fast-food workers, for instance) getting headaches when exposed to leaking microwave ovens.

?

Why does shaking a can of coffee cause the
larger grains to move to the surface?

ANSWERED BY:
Heinrich M. Jaeger, Professor of Physics,
University of Chicago.

The phenomenon by which large grains move up and small ones move down when shaking a can is called granular size separation. It is often referred to as "the Brazil nut effect," since the same result occurs in a shaken can of mixed nuts. There are several physical mechanisms that can give rise to size separation. Obviously, the very finest (dust-like) grains might just fall down through the cracks left between the larger particles. The more interesting case concerns mixtures of particles that do not differ all that much in size, perhaps by as little as 10 percent. Surprisingly, in this case the larger (and thus heavier) ones still end up near the top.

Two main mechanisms give rise to this separation. First, if during a shaking cycle (as the material lifts off the bottom of the can and then collides with the bottom again) the large particles briefly separate from the surrounding smaller ones and leave a gap underneath, small grains can move into this opening. When the shaking cycle is finished the large particles are prevented from getting back to their original positions. Thus, the bigger particles are slowly "ratcheted" upward.

Second, the granular material rubs against the side walls of the can when it is shaken. This friction causes a net downward motion of grains along the walls. This downward flow occurs in a narrow region only a few particle diameters wide close to the walls. In the center of the can, meanwhile, the particles move up, completing a convection roll. Large grains, just like any other grains, are moved up along the center (similar to being on an escalator). Once they reach the top they move toward the side wall and try to enter the downward flow. But if they are too large, they cannot fit into the narrow region that contains the downward flow and they get stuck near the top. After a few shaking cycles, this leads to an enrichment of large particles near the top.

Both mechanisms can apply simultaneously, in principle, and both will lead to the same net effect: Large particles will end up near the top. Differences between the two mechanisms are somewhat subtle. For example, the speed with which the larger particles rise to the top surface is different in the two scenarios. In practice, the second mechanism, known as a convective mechanism, dominates the first mechanism as long as the side walls are not frictionless (which is hard to achieve) and as long as you are considering particles that are not too deep below the surface.

?

Why does a shaken soda fizz more than an unshaken one?

ANSWERED BY:
Chuck Wight, Chemist, University of Utah, Salt Lake City.

Cans of carbonated soft drinks contain carbon dioxide under pressure so that the gas dissolves in the liquid drink. If the liquid is handled gently, it takes a long time for the dissolved gas to escape. If the can is shaken, however, or if the liquid is poured quickly into a glass, then the bubbles formed by turbulence provide an easier way for the dissolved gas to escape. Once the can is opened, all of the gas will eventually escape from the liquid as bubbles, and the soda will go "flat."

It's difficult for the gas to escape from an undisturbed liquid because of the liquid's surface tension, which is the energy required to separate the liquid molecules from one another as a bubble forms. For a tiny bubble just getting started, the amount of energy required per molecule of gas in the bubble is relatively large. So getting started is the difficult stage. Once it is formed, however, a smaller amount of energy (again on a per molecule basis) is needed for additional liquid molecules to vaporize and expand the bubble. The basic reason for this dependence on bubble size is that whereas the volume of the bubble is proportional to the number of molecules inside (at constant pressure), the surface area of the bubble is proportional to the number of molecules to the two-thirds power.

Because shaking the can introduces lots of small bubbles into the liquid, the dissolved gas can more easily vaporize by joining existing bubbles rather than forming new ones. By avoiding the difficult step of bubble formation, the gas can escape more quickly from shaken soda, thus resulting in more fizz.

Bottom of the 9th, Bases Loaded
the physics of baseball

?

What makes a knuckleball appear to flutter?

ANSWERED BY:
Porter Johnson, Professor of Physics, Illinois Institute of Technology, Chicago.

Actually a good knuckleball does have a slight rotation—it makes between one-half to one complete revolution in traveling from pitcher to homeplate. It is crucial to have the proper orientation of the seams on the ball as the ball travels through the air. Otherwise, the pitch would simply be a "slow fastball," which would be very easy to hit.

The principal difference between such a "batting practice fastball" and a knuckleball is that the fastball rotates several times in going from pitcher to homeplate. For a knuckleball, the important thing is that the ball rotate about an axis so that the seams are on one side of the front of the ball at one instant, whereas a little later they are on the other side of the front of the ball. The ball will then drift in the direction of the leading seam and then drift back when the seam becomes exposed on the other side.

The seams produce turbulence in the air flowing around the ball, disturbing the air layer traveling with the ball and thereby producing a force on the ball. As the ball slowly rotates, this force changes, causing the ball to "flutter" and slowly drift. The knuckleball is very difficult to throw well and is sensitive to wind, temperature, and, of course, atmospheric pressure.

In my opinion, all the great knuckleball pitchers are retired (Jim Konstanty, Wilbur Wood, Hoyt Wilhelm, Phil Niekro), and no current major league pitcher is a "knuckleballer" in the classic sense, although a few throw it every now and then. Throwing the knuckleball badly will most likely result in a so-called batting practice homerun.

?

Why does a ball go farther when hit with an aluminum bat?

ANSWERED BY:
Asif Shakur, Chair, Department of Physics and
Engineering, Salisbury State University,
Salisbury, Maryland.

A luminum is a hard material. It doesn't have a lot of "give." Therefore, very little of the ball's initial kinetic energy (the energy associated with motion) is used up in permanently deforming the aluminum. Indeed, the aluminum springs back quickly and the ball retains much of its initial energy. In contrast, wood is less elastic: It is deformed permanently and to a greater extent than aluminum. As a result, a ball colliding with a wooden bat loses more of its initial kinetic energy. At the extreme, a ball colliding with silly putty—a plastic that is completely inelastic—could lose almost all of its kinetic energy.

During an elastic collision, a ball experiences an incredibly large force for an incredibly short time, causing it to reverse direction at a speed that can be greater than its initial speed. For example, a bullet gains speed when it ricochets off an approaching artillery shell, but it loses almost all of its kinetic energy when shot into a wooden block. One must be very careful to distinguish between the expressions "losing kinetic energy" and "losing energy." The total energy is not lost; the kinetic energy is transformed into other forms of energy such as heat. Although heat is a wonderful thing on a cold morning, it does not make our ball move any faster.

Index

skin
- cancer of, 210
- fingerprints, 161–62
- skin cell replacement, 124–25
- spicy foods and, 148–50
- sunburn and, 134–35
- sunscreen protection, 210–12
- wrinkling of fingers, in water, 122–23

sleep
- deprivation, 103–5
- evolution of, 76–77
- marine mammals and, 57–60

Small, Meredith F., 89
small-world problem, 194
smallpox, 90
smart representation, 202
smell, 111–12
smiling, 99–100
Smith, Collier N., 216, 237

snow
- snowflake symmetry, 167–68
- temperature and, 170

Sobell, Jeffrey M., 134
sonic boom, 221–22
Sottery, John, 210

sound, 216–22
- booming sands, 219–21
- sonic boom, 221–22
- from voltage power lines, 217–18
- whip "crack," 216–17

space travel, 2, 3
Spacewatch Telescope, 1
Spagna, George F., Jr., 13, 19
Spatola, Arno F., 155

speed
- of earth's movement, 17–19
- measuring with laser gun, 197–99
- of sound, 216–17
- spinning propeller illusion, 237–38

SPF, 211
Spider-Man (movie), 46
spiders, 45–46
spiracles, 42
squids, 47–49
stainless steel, 141–42

stars
- galactic rotation, 228–30
- globular clusters, 27–28
- Leo constellation, 19
- life expectancy of, 23–24
- North Star, 21–22
- rotation curve, 229
- Sagittarius constellation, 28
- the sun, 9
- twinkling appearance of, 25–26
- Ursa Minor constellation, 21